THE DEEP STATE

THE
DEEP STATE

**How an Army of
Bureaucrats Protected
Barack Obama and Is Working
to Destroy the Trump Agenda**

JASON CHAFFETZ

BROADSIDE BOOKS
An Imprint of HarperCollins*Publishers*

HarperCollins books may be purchased for educational, business, or sales promotional use. For information, please email the Special Markets Department at SPsales@harpercollins.com.

FIRST EDITION

Photograph by nicemonkey/Shutterstock, Inc.

Library of Congress Cataloging-in-Publication Data has been applied for.

ISBN 978-0-06-285156-7

18 19 20 21 22 LSC 10 9 8 7 6 5 4 3 2 1

Dedicated to the American people. May they always remain vigilant to keep their government open, transparent, and accountable.

Contents

Introduction

I t is October 12, 2012, and I am in Stuttgart, Germany, on my way to Libya. I am standing in the office of four-star U.S. general Carter Ham, about to get a classified briefing on everything the United States knows about the attack less than four weeks ago on our diplomatic facility in Benghazi, Libya. I am about to find out, I hope, why four Americans, including our ambassador, died during the attack and why no nearby American forces were deployed to save them. I am a U.S. congressman and I have top-secret security clearance.

No one introduces me to a young, blond-haired man in the room holding a notebook and pen. I do notice that people in the room seem a little intimidated, a little too aware of this fellow. So I ask who he is and that I be introduced.

He is a lawyer from the State Department who has just arrived from Washington, D.C. He knows nothing about Libya. He is not a specialist in anything remotely having to do with the country or terrorism or the military or conflict zones.

He will contribute nothing. He is a State Department lawyer who specializes in Freedom of Information Act requests—or, more specifically, how to keep information hidden from the public . . . and from a congressman.

He next appears at a classified briefing in Tripoli. I have had it with being watched over by a State Department "minder." I want him out of the Tripoli meeting. He refuses. He wants to be in the room when I am briefed. He telephones Secretary of State Hillary Clinton's chief of staff, Cheryl Mills. Now there is a standoff. I want him out.

Well, it turns out he doesn't have the proper security clearance for this meeting.

He is out.

This story—this book—is *not* about Benghazi. What happened at Benghazi, and more specifically, the story I've related here about my attempt to uncover the truth about Benghazi, is ultimately about the very much alive and thriving Deep State. It is about control of information to the American people and the control of the truth.

That moment, facing the young blond-haired State Department lawyer, was the first time I knew I was face-to-face with the Deep State.

Unfortunately, it was not my last confrontation with them.

The Deep State is real. They don't like exposure, accountability, or responsibility. They fight back, outlast, and work the system for their advantage. And they certainly don't like disruptive forces such as Donald Trump. For example, but for a few brave souls—whistleblowers who willingly put their careers on the line—the absolute duplicity of Hillary Clinton's team and the Obama administration's response to the Benghazi attacks would never have been exposed.

This is a book about what happens when huge swaths of government begin prioritizing their careers over getting the job done. Many readers will think they know this story—a story of incompetent drones who are underqualified and lazy. When we talk about the Deep State, we get this story all wrong. We misunderstand what has happened, and mischaracterize who is doing it.

Sometimes even conservatives talk about the Deep State as though the term refers to dumb, inefficient bureaucracy.

In fact, it is the opposite: the Deep State is intentional, unconstitutional, and organized. It is about pure, unfettered power, and it gets very angry when it is even questioned.

Their stonewalling, their inability to coherently defend themselves, their secrecy—these look like weakness, but they are the Deep State's greatest strength. We need a better guide to the parts of the federal government that are actively working against the will of the people.

The Deep State has an agenda. It rarely matches up with conservative principles.

The stories in this book are always frustrating and sometimes shocking, but they will challenge many closely held assumptions about the Deep State. The heart of this book, though, is unraveling the greatest puzzle of the Deep State: Congress rarely does anything about it. As chairman of the House Committee on Oversight and Government Reform, I was the tip of the spear challenging the Deep State and trying to hold them accountable. In this book I highlight their tactics, illuminate the problems, and offer a way to fight back and win. It is important to expose the stories, but if the American people are going to win, Congress is going to have to do things differently, and this book helps concerned citizens understand so they can engage.

The Deep State has been in place for a long time, arguably since the middle of the twentieth century. As a new member of Congress, I got to see it in action as we took on the powerful forces at the Internal Revenue Service, the Environmental Protection Agency, the Department of Justice, the Department of State, etc. The deeper I dove in, the more shocking was the brazen approach of the power brokers. They were used to operating anonymously and without consequence. This problem is bigger than we can imagine and getting worse, unless we do something dramatic to wrest back control.

Protecting Big Government

Washington, D.C., will hand out more than $4.3 trillion in the next twelve months, or roughly $12 billion per day (including about $700 million per day in interest on the national debt). What could go wrong?

I believe the United States of America is exceptional and inspired. I also believe that after 240 years we have an obligation on our watch to provide diligent stewardship. We must pass our country to the future generations better than ever.

Responsible parents do all they can to create a better life for their children. This includes leaving our country better than we found it.

The United States of America is the greatest country on the face of the planet, but we have a responsibility as citizens to engage in the management of our government, our resources, and the American people's money. Government does some good things, but it can also ruin our lives or reach too far into them.

Our country's long-term success includes living within our financial boundaries, limiting government to the powers envisioned by our Founders, defending ourselves with a dominant military, and ensuring accountability within the omnipresent government apparatus.

Not everyone sees it that way. The problem: many of those who disagree with everything I've just said work in the government.

Yes, the Deep State is real. It's been characterized as different things by different people. Some call it the government's massive secrecy apparatus as conducted by the federal defense and intelligence agencies; others include a handful of federal trial courts, corporations, and private interests, including Wall Street and big banks, in their description of the Deep State.

All of that may be true. I tend not to imagine nefarious conspiracies in dusty dark corners. I tend to see ordinary imperfect human beings and ordinary human institutions, acting too often with greed, fear, and with messy self-interests. The Deep State is not Democrat or Republican. It doesn't wear a trench coat, and it doesn't linger furtively on street corners at midnight in Washington, D.C. In truth, and perhaps most startling, the Deep State doesn't hide at all.

In fact, the reality is much worse than that. The Deep State is a vast, self-perpetuating bureaucracy whose aim is singular: to exist again tomorrow and the day after, to replicate itself, to be indestructible and nearly impossible to disrupt. As a congressman, I have seen it and experienced it up close in my confrontations with the State Department, the IRS, the EPA, the Secret Service, and the Justice Department, just to name a few. U.S. presidents come and go, political parties win one election cycle and lose the next. The Deep State does what it wants, and waits out periodic blips. Inevitably, it seems, the Deep State goes on.

Could the bipartisan Deep State live with a President Jeb Bush? Absolutely. President Hillary Clinton? Even better. I could throw out a bunch of other acceptable figures who ran for president in 2016 and in many past elections.

But President Donald Trump? No. That man was an anomaly whose election could simply not happen. And yet, of course, it did. Now, Washington has lived for a long time with the occasional political outsider who came to Washington, even some who did not conform to the ways of Washington—think Jimmy Carter, perhaps. Democracy does that sort of thing from time to time. But Donald Trump was more than an outsider. He campaigned not just against Washington, as most politicians do, but against the New York–based media and the coastal elites. Worse, the problem was that Trump was a traitor! He was not some peanut farmer from Georgia. He was a billionaire New Yorker who knew the media more than anyone, had traveled in their circles, been invited to their parties and private clubs and invited the elites to his. Who can forget the photograph of Hillary Clinton beaming at Donald Trump at his wedding to Melania? And yet here he turned on them. A man who loved America so much he was not going to "play the game." His constituency, the voters who filled stadiums throughout the primary and general election season, was, as Hillary Clinton memorably called them, the "Deplorables," the working-class middle Americans, forgotten men and women left behind by crony capitalism, unbridled globalization, illegal immigration—and the Republican and Democratic establishments.

What we know about the Deep State is admittedly the tip of the iceberg. Even that tip would require more than one book to adequately explore. With more being revealed every day, writing the full and conclusive story of the Deep State—even just within the Obama administration—would be almost impossible. The story is a moving target. This book will contrast my experience with the Obama administration during the eight years I served in the United States Congress with what we're seeing play out today under President Trump.

Trump, the ultimate disrupter and rule breaker, won an election he was simply not supposed to win. And the Deep State, a wounded bear, is having none of it. Before Trump was even sworn into office the Deep State went into overdrive to thwart his presidency. In agency after agency, the threat to big government posed by Trump changed all the rules. At the top of the chopping block was the Obama-era Consumer Financial Protection Bureau.

CHAPTER 1

Elizabeth Warren's Big Dream: The CFPB

E lizabeth Warren had a dream. Unlike many of us, she did not dream of a new house, a newer car, a higher-paying job, or even world peace. Her dream, which came true, was for a new federal agency with 1,623 employees and an annual budget of $605.9 million.

In 2007, she wrote an article titled "Unsafe at Any Rate" for the statist publication *Democracy Journal*. The article called for the creation of an agency almost identical to the Consumer Financial Protection Bureau.

It's an interesting article. While it smartly discusses needed protection for consumers, it also goes further in its tone. It outlines a sharp perspective on divisions in America—between rich and poor, dependency and personal responsibility. Like so much leftist writing, the article's tone is divisive and essentially declares that the wealthy are evil and all-powerful, and that the poor and middle class are helpless and without agency for their own lives:

Indeed, the pain imposed by a dangerous credit product is even more insidious than that inflicted by a malfunctioning kitchen appliance. If toasters are dangerous, they may burn down the homes of rich people or poor people, college graduates or high-school dropouts. But credit products are not nearly so egalitarian. Wealthy families can ignore the tricks and traps associated with credit card debt, secure in the knowledge that they won't need to turn to credit to get through a rough patch.

Comparing credit card debt to a deadly malfunctioning appliance seems weird to me . . . but okay. Warren was a law professor at the time but soon ran for office and was elected a U.S. senator from Massachusetts in 2011.

She then eagerly set about preventing the next financial crisis with legislation creating a bureaucracy that had very little to do with the causes of the previous one.

The Dodd-Frank Act describes the CFPB's mission:

The CFPB was created [in July 2010] to provide a single point of accountability for enforcing federal consumer financial laws and protecting consumers in the financial marketplace. Before, that responsibility was divided among several agencies. Today, it's our primary focus.

   Our work includes:

- *Rooting out unfair, deceptive, or abusive acts or practices by writing rules, supervising companies, and enforcing the laws that outlaw discrimination in consumer finance*
- *Taking consumer complaints*
- *Enhancing financial education*
- *Researching the consumer experience of using financial products*

- *Monitoring financial markets for new risks to consumers*
- *Enforcing laws and monitoring risk.*

Who could be against that?

The Reality of the CFPB

If the Deep State could wrest control of government away from the people and design their version of the perfect government entity, they would probably design something like the Consumer Financial Protection Bureau. This Obama-era agency checks every box on the Deep State's wish list: unlimited, unaccountable, and unchecked. The funding is nonrescindable, the leadership supposedly untouchable, the decisions unquestionable, and the operation impervious to the political winds. Power, money, and no accountability. What more could the Deep State ask for?

Ironically, Democrats sold the agency as a solution to the very problems it would create. Representative Barney Frank of Massachusetts and Senator Christopher Dodd of Connecticut said the CFPB was intended to consolidate regulation and hold people accountable.

So they created an agency of unaccountable people to write more regulations. If that doesn't make sense to you, you are probably not a member of Congress.

They created an entity with no checks and balances to check and balance the financial markets. They hired bureaucrats to operate in secret with the mission of shedding light. They hired political hacks to take the politics out of regulatory oversight.

Sold as a way to consolidate regulation, the Dodd-Frank reforms didn't consolidate anything. As only Washington can do, the legislation grew the bureaucracy, expanded regulations, and spent millions more of your dollars. This agency is the epitome of the progressive agenda: more government, more regulation, more of your money, more protection from yourself, and less liberty.

I doubt there is a more egregious example of putting government out of reach of the people than this monstrosity of government waste and abuse. This is one more case of the federal government spending millions to educate us about what is best for us.

In addition to its 1,600-plus employees, the bureau is partially housed in the heart of Washington, D.C. You, the taxpayer, spent more than $139 million on a building that was more than 300 percent over budget to make sure they have only the finest of offices in the highest rent district possible. The agency spends hundreds of millions of dollars, but it is shrouded in secrecy and lacks the basic transparency found in other parts of the federal government.

Warren appeared before the Oversight Committee weeks before the bureau opened for business in 2011. The House Committee on Appropriations had cited several instances where they had little to no insight into the CFPB's budget numbers. I tried to question Warren in my five minutes about the budgeting items of the CFPB but she didn't have a clue, or if she did she stonewalled with ignorance.

This was a newly established bureau whose mission was to root out deceptive practices and provide a clearer light into the inner workings of financial institutions. However, its own operations and numbers were lost in a cloud of secrecy.

Most unsettling is the mechanism for funding the bureau, which is extremely unusual and little noticed outside Washington.

Under our normal federal system, government agencies are clearly within the bounds of the executive branch, but their funding comes via Congress. Almost every agency's funding comes through congressional appropriations. The House originates all spending bills, differences are worked out with the Senate, and eventually the bill makes its way to the president for signature.

But not the Consumer Financial Protection Bureau.

The CFPB was purposely designed to bypass Congress, checks and balances, and oversight. It is funded by the Federal Reserve, also known as the Fed, and therefore is outside the reach of the United States Congress. All funding per the Constitution is supposed to originate in the House of Representatives, but the CFPB's appropriations don't happen this way. Those monies are allocated by the Fed, but the public has no true insight into the Fed's financial operations.

Don't for one minute think that funding from the Fed means you aren't paying for it. The Fed generates revenue by collecting interest on the $4.5 trillion in bonds it owns. Where does the Fed get the money to buy all those bonds? Thin air. Not even joking. It's a neat trick. They simply create new dollars that didn't exist before. In doing so, they devalue every dollar you and I hold. Then they collect interest on the bonds they bought and use that interest to fund their operations. What's left over, they pay into the Treasury. The more they spend on the CFPB, the less they pay to the Treasury. You see that sleight of hand? Then the money comes out of the federal budget, but not in a way Congress can control in the future.

What is the CFPB spending? We don't know. Information about the Federal Reserve's financial operations is fiercely protected by the Deep State.

I'm not exaggerating when I say this is close to a black ops intelligence operation.

After I first arrived in Congress in 2009, I ended up cosponsoring a bill with Congressman Ron Paul of Texas. His bill would require an audit of the Federal Reserve. In fact, the bill was often referred to as "Audit the Fed." It seems simple enough. It seems fair. It was wildly popular back home and across the country. Even Barney Frank was in favor of the Audit the Fed bill. In fact, he cosponsored it.

It would not go on to become law.

That failure was a classic example of a good idea with broad support that still went nowhere. Congressional "leadership" worked to make sure that bill never actually got to the president's desk. So many members wanted to tell the voters back home that they were in favor and cosponsoring the bill, but somehow the bill would never pass both the House and Senate in the same Congress.

There is a little-known process to bring a bill up for a vote on the floor of the House of Representatives without the consent of leadership. It is called a discharge petition. If a discharge petition is introduced with simple paperwork on the floor and a majority of members in the House sign it, the bill must immediately be brought up for a vote.

Leadership hates this. Through the years I served in Congress it was a little-used process, except during the reauthorization of the Export-Import Bank of the United States. (I opposed that.)

With 435 members in the House of Representatives, the process requires only 218 members to win a vote and successfully use the discharge petition to bring that bill to the floor. Most members in the majority are reluctant to publicly sign the document, since leadership views it as an embarrassment. But the Speaker or majority leader would secretly and verbally give members the okay to sign the petition if it helped them back home, as long as they were not the one who pushed it over the 218 mark.

Ron Paul's bill had more than 300 cosponsors. He eventually introduced the discharge petition. Unfortunately, we never got more than 218 people to sign the petition. Nearly a third of the bill's supporters wanted the credit for cosponsoring the bill without having to vote for it. The bill eventually got a vote, but it didn't pass the Senate. The fight continues.

Meanwhile, the CFPB has a budget that is reportedly bigger than the Securities and Exchange Commission (SEC). Its mission is so broad there are no bounds to its areas of jurisdiction. It is an agency subject to very little oversight.

Many times people will suggest Congress should use the so-called power of the purse to check and balance an out-of-control agency like the CFPB.

One problem: there is no use of the "power of the purse" when an agency is funded by the Federal Reserve. Congress has a hard time even getting a response from the CFPB, let alone holding them accountable. Their budget is mysterious and lacks detail by design. Yet CFPB is one of the larger agencies in the federal government, particularly as it relates to regulations.

Congressmen Jeb Hensarling and Bill Huizenga have fought hard to try to dismantle the CFPB, but there's no way the Democrats want to let go of that one. It should scare all of us to have a massive government regulatory body in D.C. with the power to regulate extensively while beyond the scope of Congress. Because the CFPB doesn't get its funding from Congress, it doesn't feel accountable to it. And that should scare all of us. Congress is a body made up of representatives of the people. The CFPB doesn't answer to the people. It answers to the Fed.

The CFPB is particularly potent on publicly traded companies. As publicly traded companies, these entities must disclose the mere existence of threats. I have heard several examples in which the CFPB will make its presence known in a threatening posture and offer "guidance." These aren't laws, nor are they even rules that have run the gauntlet of review and comment. "Guidance" is another way of saying you're going to do what we tell you to do or else your life will become very difficult and financially painful.

Honestly, those are reminiscent of mob intimidation tactics.

The CFPB has the ability to issue a press release. Press releases can devastate a publicly traded company and literally shed millions of dollars of market capitalization from its valuation in a matter of minutes.

It generally works like this: Via a nameless bureaucrat the CFPB or other regulators will find an entity or business model it does not like. They will begin an investigation and make themselves known to that company and eventually offer "guidance." The company then has a very difficult choice: they can implement that guidance or they can choose due process. But this is not a fair fight, as there is no recourse or due process. The CFPB holds all the cards.

Companies are routinely held hostage without the ability to seek recourse or petition their government.

This CFPB behavior might seem like a good idea when the company is wrong. The company is publicly shamed for their bad behavior, and the whole industry is on notice that they have to do better. But what happens if the company isn't wrong? What if the company has been unfairly targeted? What if it's just a mistake?

The answer for the company is surrender anyway or else face a nearly impossible battle in court and the public square. Because Congress can't do much to help it.

Protecting Itself Becomes the Mission

You may be thinking, or hoping, even, that the CFPB is actually a nonpartisan good-government agency concerned with protecting consumers.

I'm sorry. Prepare to be disillusioned. It gets worse. Cronyism and corruption quickly engulfed the agency.

President Obama famously used an advertising agency called GMMB, headed by Democratic strategist Jim Margolis. President Obama paid them more than $700 million for his 2008 and 2012 campaigns. Margolis also worked for Bill Clinton and for Hillary Clinton in 2016.

Apparently the CFPB also needed good public relations and advertising. The CFPB has paid GMMB $43 million. Even after President Trump took office in January 2017, the CFPB has paid Obama's ad agency $15 million. What was that money for? "Advertising placement, media planning, media buying, consumer research, creative development and creative testing," according to the contract. One of the purchase orders, for $950,474.01, was "to develop a CFPB marketing strategy and the Owning a Home product." This was described as an online tool only. An online tool to teach you how to buy a house, find out what current mortgage rates are, and get a mortgage.

Let's put aside that the free market already provides this service. Most people looking to buy a house may stop by the local Realtor's office for listings, or look at Zillow, or google "mortgage rates." This arrangement is the kind of government corruption and cronyism that would be funny if it weren't coming out of our pockets.

We will see this over and over again throughout this book: a government agency spending obscene amounts to educate the public on things people already know.

And to be clear; this isn't just government "waste." This is government corruption—taxpayer dollars going into the hands of political consultants.

More drama was to come following President Trump's inauguration.

Richard Cordray, the controversial director of the CFPB, decided to step down while Donald Trump was the president. Cordray's term was set to expire in 2018 and surely he would not be renominated.

Tucked into the legislation that created the CFPB was a little-known provision that supposedly allowed the deputy director to automatically become the acting director of the CFPB even without the president's consent. Liberals were counting on using this provision to maintain control of the CFPB. They weren't counting on President Trump fighting back.

Upon Cordray's stepping down, his deputy, Leandra English, assumed she would show up for work the next day to continue to lead the agency.

President Trump saw it differently. Citing the Federal Vacancies Reform Act, the Trump administration defended the president's right to appoint someone to a vacancy that normally requires Senate confirmation. Under this act, the president can fill a vacancy with another person who has already been confirmed by that body.

President Trump selected Office of Management and Budget director Mick Mulvaney.

Mulvaney was a former member of the House from South Carolina. He sat on the Oversight and Government Reform Committee while I was the chairman. He is a fiery Irishman with a mind for numbers. He could be exceptional in questioning administration officials. That experience paid off in a big way. The OMB director is frequently on Capitol Hill and is bombarded with questions.

Since he joined the White House, Mulvaney's wardrobe has picked up dramatically, but he still has a penchant for the vests that routinely made him the subject of ridicule on the floor of the

House. Members tend to obnoxiously and habitually poke fun at one another, especially when wearing a vest that makes you look like a character from the game of Clue.

So, the old guard at the CFPB decided Mulvaney shouldn't take office under President Trump. When the following Monday rolled around, both Mulvaney and English showed up to work to lead the agency. Oh, how fun it would have been to see the two of them go to the director's office at the same time.

I know Mulvaney. No way was he going to back down on this one.

"It is unfortunate that Mr. Cordray decided to put his political ambition above the interests of consumers with this stunt," White House spokeswoman Sarah Huckabee Sanders was quoted as saying in the *Washington Post*. "Director Mulvaney will bring a more serious and professional approach to running the CFPB."

The presidential selection of Director Mulvaney was a clear signal that this administration wanted hands-on leadership in this dangerous agency.

The whole matter went to court and President Trump's decision prevailed. I still find it unbelievable that it took a ruling from the U.S. District Court of Appeals for the District of Columbia Circuit for a president to be able to appoint somebody to run an agency in the executive branch.

While the Trump administration was victorious in court in this round, it does concern me that Democrats could continue to use the Federal Reserve to bypass Congress. Ultimately, it means they don't trust the American people, and they don't want accountability.

President Trump is trying to cut the CFPB budget by $150 million. Let's hope he succeeds. Going further, Acting Director Mulvaney has recommended zero funding. The reason is that once a federal agency is established, especially one that is a poster child for Deep State interests like K Street lobbyists, it can be very difficult to get rid of.

That brings me to another newish federal agency that everyone agrees we need, but no one likes.

CHAPTER 2

What Don't They Want Congress to See?

The Transportation Security Administration, or the TSA, as most of us know it, was formed after the attacks of September 11, 2001. Every reasonable American would agree that those attacks on our country and the subsequent years of Islamic terrorism have made extreme protection of our country's airports and mass transportation system a top priority.

The 2017 budget for the TSA, which was made a part of the Department of Homeland Security in 2003, is $7.6 billion. The TSA employs more than 60,000 people, including 350 explosive specialists, 2,000 behavior detection specialists, federal air marshals, and explosive detection canine teams. They protect not just our airports, but also our railways and highways, cargo, tunnels, pipelines, and bridges.

Like so many federal agencies, the TSA is better at spending money than at getting results. A 95 percent failure rate on detecting hidden weapons, reported in 2015, resulted in acting TSA administrator Melvin Carraway being reassigned. The TSA screeners, in tests done by the Office of Inspector General for the TSA, failed to detect banned weapons in 67 out of 70 screenings. Similar results were found at individual airports. In a July 2017 test at Minneapolis–St. Paul, banned items got through screening in 17 out of 18 tries. The TSA refused to release the results of more recent tests, but in November 2017, ABC News reported a source had cited a failure rate in the area of 80 percent.

With those results, how much security is our $7 billion investment in the TSA really buying us? Apparently, the TSA leadership is happy enough with the results to offer massive bonuses to those at the top—even as the agency's turnover rate is among the highest in the federal workforce. While the attrition rate for the federal workforce as a whole is a low 6 percent, the TSA's rate is 9.5 percent. And that's the good news for the TSA. Among part-time employees, which make up 23 percent of the TSA's workforce, the turnover rate is 19 percent. The TSA is perpetually spending money to train replacements. I liken it to demanding we pour more water into a tub while the drain is open. The agency consistently ranks last on the annual survey of best places to work in the federal government.

Meanwhile, those at the top can qualify for massive bonuses. The inspector general reported that the TSA assistant administrator for the Office of Security Operations Kelly Hoggan was paid nine different bonuses in the $10,000 range between 2013 and 2014, for a total of $90,000—on top of his $181,500 annual salary. One would hope that was an extreme example—which led to Hoggan's resignation in May 2016 following a hearing before the House Oversight Committee. That hearing explored high attrition rates, poor disciplinary policies, retaliatory reassignments against disfavored employees, a lack of accountability among senior staff, and, of course, the bonuses awarded in the face of poor performance. Against that backdrop, let's take a closer look at the technology meant to detect weapons and who benefited from the purchase of that technology.

In 2008, the TSA started using body scanners in ten airports around the country, and soon announced that thirty-eight machines would be set up within weeks.

We want to be safe. But those scanners were seriously invasive of a person's privacy. Let's just say that, in addition to health concerns, those scanners would clearly show a person's sex.

Now, any passenger could decline the scanner, and ask for a pat-down instead . . . but only 4 percent of passengers did. I can understand why, as you'll soon see.

We can all agree that $90 million is serious money. That is the amount of the federal contract awarded to a company called Rapiscan, one of the most unfortunately named companies in history. Rapiscan made those first-generation airport body scanners that cost about $180,000 apiece. There were lots of potential problems with those scanners, but to me one of the biggest was that Rapiscan was a client of Michael Chertoff, through his company the Chertoff Group. Chertoff was the secretary of the Department of Homeland Security (DHS) from 2005 to 2009.

Chertoff founded his firm just after Barack Obama took office as president and he recruited at least ten top officials from DHS and the CIA. The *Huffington Post* called it a shadow homeland security agency.

This kind of mutual back-scratching just isn't right. Representative Ron Paul tried to stop the purchase. "Here's the guy who was head of the TSA, selling the equipment. And the equipment's questionable! We don't even know if it works and it may well be dangerous to our health," Paul said on the House floor.

According to the *Huffington Post*, the other company manufacturing similar scanners, L3 Systems, spent more than $1.4 million lobbying the government. Its chief lobbyist was Linda Daschle, ex-wife of former Senate Democratic majority leader Tom Daschle.

Is this whole thing starting to sound like the Swamp to you? It sure did to me.

I don't often side with the American Civil Liberties Union, but I did on the issue of personal privacy. I called the scanners "TSA porn." A measure banning them was the first piece of legislation I introduced as a congressman in 2009.

I started asking some questions and getting some briefings. That, combined with some information from whistleblowers, which I'll discuss soon, and some classified briefings made me decide to pursue the issue.

There are 5,136 airports and more than two million people who get on an airplane in the United States every day. The TSA is responsible for security at 440 airports. I was flying back and forth from Salt Lake City to Washington, D.C., every four days or so. It took a little while to figure it out, but soon it became evident the so-called body scanners were easy to beat.

From what I learned—and saw—the devices were more security theater than anything else. I'm sure there was a bit of a deterrent factor, but would-be terrorists would soon figure this out as well.

Listen, finding and detecting explosives is important, whether it's on airplanes or inside buildings, or on the ground in war zones. It's critical. But it is also big business. The Pentagon spent more than $20 billion studying how to beat improvised explosive devices, or IEDs. We were losing too many Americans in Afghanistan and Iraq to them, so the Department of Defense asked a general to head up what was first known as JIEDDO, or the Joint Improvised Explosive Device Defeat Organization. After spending $20 billion, their conclusion was—are you ready?—the single best tool to detect IEDs is a dog. Yes, a dog.

But dogs don't have lobbyists. That's why you see more machines than you do dogs. Even though it's more effective and less expensive, the TSA was succumbing to the billions of dollars appropriated by Congress. Like every federal agency in the bureaucracy, they felt the need to spend the money allocated to them.

Combine that Deep State impulse with lobbyists and you have a formula for expensive things that don't work, something we saw in the last chapter as well.

Lobbyists became involved and began selling machines that were less effective than canines. The dogs are friendly. They're mobile. They scare terrorists because the latter don't know what the dog is sniffing for—bombs, drugs, cash, or something else. You can use them everywhere from the parking lot all the way to the plane. And they are amazingly effective.

I once went to Afghanistan, visiting Camp Leatherneck. The plane that landed just before unloaded kennels for dogs. I had my picture taken with them. Jokingly I asked the guys, "Hey—where are those WBI machines?" (WBI stands for "whole-body imaging.") What? "How come you don't import those WBI machines to find these IEDs?" When I explained, they started laughing.

"This is life-or-death. We're not going to use some stupid machines that we know how to beat. We like our German shepherds," they said.

You don't see these machines at the White House. Or on Capitol Hill. They would never use them at the White House because they don't work. When it's serious, they don't use them.

The Conflict Gets Personal

A *USA Today* story I read highlighted the fact that these machines could see the bead of sweat on your back. I insisted that the TSA show me how these machines work. I wanted to go behind the scenes and see people getting scanned.

Remarkably, the TSA refused to do it. I pushed and pushed and pushed. They said that for privacy reasons they couldn't let me look at these machines. I said, Why not? In the back room, a TSA person is sitting at a large television monitor. When they finally let me go back in that room in Salt Lake City, they weren't doing any scanning. They just showed me where the screen was. This was hardly showing me how the screening worked and what level of invasiveness it entailed. I did the same thing at Reagan National Airport in Washington, D.C., to see if this obstruction was just a Salt Lake City problem.

It wasn't.

In both circumstances, the TSA legislative liaisons became, shall we say, highly engaged. A number of them flew out to Utah so they could be there to supervise me looking at this process. But they wouldn't let me witness a live session where people were being scanned. That was a flashing red light. They said that because of privacy concerns I could not watch someone doing their job.

Think about that. The TSA was willing to allow regular TSA employees to look at countless people via this machine and they claimed there was no problem. But it was too sensitive for a member of Congress on the Oversight Committee to look at the scans of anonymous people because of the privacy concerns. I had a top-secret security clearance and was elected to do oversight, but that wasn't good enough for Homeland Security.

This is a theme we will return to again and again in the book: departments and agencies doing their very best to prevent a member of Congress from coming in and looking around.

When I spoke privately to the TSA personnel, they told me, "If I really wanted to, I could read the date on a coin." If that was true, that would show that these people could look at men and women in the most intimate ways. (If you think this is a silly concern, you will be especially interested in a later chapter covering Deep State employees spending hours every day downloading pornography.)

The TSA was insistent that no images were taken or transmitted. That is a lie. I finally obtained the actual request for proposal, as such are known, from the TSA to the potential manufacturer. The machines are built with USB ports and storage capacity, and obviously they are transmitting what was taken as a picture and sent to another room for analysis! The machine was built to transmit photos. Even though they testified publicly, under oath, it was a flat-out lie. When I got too technical in my questions, they feigned ignorance.

Media reports also indicated that U.S. marshals had ten thousand of these images stored in Florida. Word got out pretty quickly, particularly in Salt Lake City, that I was a problem for the TSA. Despite my being a frequent flyer and a member of Congress, it became more than a coincidence that I was routinely pulled out of line and asked to go in the scanning machine, which I refused. That meant I had to have a very invasive pat-down.

I don't know that I've ever had such aggressive pat-downs. But it was part of the price the Deep State wanted me to pay for looking under the hood.

In September 2009, I had a major run-in with the TSA at the Salt Lake airport. I'd gotten in line to go through a metal detector. But when I got to the front, I was directed to go to the machine scanner. I refused. I'd gotten in line for the metal detector.

In truth, other than being treated rudely and confronted by a TSA supervisor who refused to give me his badge number, nothing happened. A TSA officer later nastily claimed that I had said things and acted exasperated. Fortunately, Thomas Burr from the *Salt Lake Tribune* asked for the video through a Freedom of Information Act request. It took weeks to come out but the video backed up my story.

The union representing TSA employees got involved, criticizing me even before they had seen the video. Interesting.

The good that came from my aggressively questioning the TSA tactics and investment in these machines led to a huge change where an algorithm would produce a stick figure that would highlight anomalies for further consideration. It changed the equation dramatically. The drama and various hearings continued throughout 2011 and 2012. Despite assurances that TSA representatives would testify, they often backed out at the last minute.

The FAA Modernization and Reform Act, passed in 2012, required that all full-body scanners operated in airports use "automated target recognition" software, which replaces the picture of a nude body with a cartoonlike representation. As a result of this law, all backscatter X-ray machines formerly in use by the TSA were removed from airports by June 1, 2013.

The TSA said it was because Rapiscan, the backscatter machine manufacturer, did not meet their contractual deadline to implement the software.

The TSA opted instead for "millimeter-wave" scanners. Besides being less invasive and generating fewer health concerns, the TSA claims millimeter-wave scanners also move passengers through faster.

That software also allows for more privacy by generating the images with a generic outline, not the "nude" outline I and others objected to. According to *Wired* magazine, "Rapiscan came under suspicion for possibly manipulating tests on the privacy software [the automated target recognition software] designed to prevent the machines from producing graphic body images."

By 2016, the Department of Homeland Security issued a 157-

page final report on passenger screening using Advanced Imaging Technology (AIT). DHS stated that AIT is the most effective way to screen passengers. AIT is what they're calling the full-body imaging technology now.

The story illuminates how the Deep State is not bashful in pushing back against Congress. And today? Though the TSA claimed to up their investment in canines, they are now claiming there is a shortage of bomb-sniffing dogs.

You can't make this stuff up.

Money, Sex, and the EPA

The competition for worst-managed agency in the federal government is pretty stiff, as you might imagine. But in my opinion, the winner hands down is the Environmental Protection Agency (EPA). It's not even close. Words like *overreach* don't even begin to describe the abuses perpetrated by this rogue agency. The level of mismanagement we saw during the Obama years is a classic example of what happens when accountability is nonexistent. Without oversight, the agency itself became toxic. Even as people and business are suffering from the abuses, the agency steadfastly resists further scrutiny.

But as you will see, those who are part of the Deep State don't have to worry about incompetence or mismanagement. They pay no consequences for either.

The EPA was formed in 1970 to safeguard the cleanliness of our air and water and ensure safety in the use of chemicals. Today it is an agency with about 15,000 employees and a budget of $6.14 billion. (And that's *after* President Trump has called for a 23 percent budget reduction.)

Here's one small example of the kind of EPA program President Trump is cutting:

REDUCE RISKS FROM INDOOR AIR (FY 2018 ANNUALIZED CR: $13.386 M, 40.7 FTE)

This program addresses indoor environmental asthma triggers, such as secondhand smoke, dust mites, mold, cockroaches and other pests, household pets, and combustion byproducts through a variety of outreach, education, training and guidance activities. This is a mature program where states have technical capacity to continue this work.

Yes. You read that right. President Trump is saving taxpayers $13 million by eliminating a program that seeks to educate people on how to cope with cat hair and cockroaches.

Now let's talk about former EPA administrator Gina McCarthy. Before she became the EPA top dog, she was assistant administrator from 2009 to 2013. So, let me share a story about somebody she directly supervised. I'll tell you right now, I'm not going to be able to resist giving you all the details of this case. It's that outrageous.

You can judge for yourself.

Did You Hear the One About John C. Beale?

He's not a Super Bowl star or the newest winner of *The Voice*, nor is he dating a Kardashian. He's not even a politician. No, John C. Beale was once the highest-paid employee at the EPA making $206,000 a year.

And, as it turned out, John Beale defrauded the U.S. government out of $886,000.

How did he do it? Well, mainly by claiming to be a CIA spy.

When he appeared before the House Oversight Committee on

October 1, 2013, I told Beale that his tenure with the EPA smelled a lot like the Leonardo DiCaprio movie *Catch Me If You Can*.

Beale is originally from St. Louis County, Minnesota, which is about an hour and a half from the county seat, Duluth. It's a Democratic stronghold where mining and forestry are important industries. John Charles Beale was born in 1948. His mother, Marcella, was a nurse and his father, Charles Gordon Beale, was a minister. The family moved from Minnesota to Greenwich, Connecticut, and then to Bakersfield, California, when the senior Beale accepted new ministerial positions.

Young Beale graduated from Bakersfield High School in 1967, went to Chapman College in Orange County for two and a half years, then left school to become a police officer in Costa Mesa. He went into the army in 1971 and trained as a physical therapist and a medic.

After an honorable discharge in 1973, a 2013 Senate report said, Beale led an "itinerant" life in California, going to school at the University of California, Riverside, and earning a degree in political science. As a senior, he briefly interned for Democratic U.S. senator John Tunney. Beale characterized himself as "a very nomadic type of person" in his 263-page deposition to our committee on December 19, 2013.

In 1975 he went to law school at New York University and from there it was off to graduate school at Princeton's Woodrow Wilson School of Public and International Affairs. There he met Robert Brenner, later a deputy assistant administrator at the EPA. That was a very important connection, as we'll see. After briefly working at a Seattle law firm, Beale returned to Minnesota to work on his cousin's apple farm. Just before joining the EPA, Beale was employed at a three-person law firm in Lake City, Minnesota.

Brenner and Beale, by now good friends, bought a rental property in Massachusetts from Beale's parents in 1983.

With no formal training or experience in EPA matters, Beale became a consultant to the agency in 1987, hired by Brenner. The job, he recalled in his deposition, would "give me a lot of work with the Hill, which I'd had some experience in, would require a lot of liaison work with the Office of General Counsel, because that's inevitable when you're drafting legislation, and as an attorney, I knew how to talk to lawyers."

When asked if he and Brenner had discussed the possibility that his lack of environmental experience might impede his getting the job, Beale said, they had talked about it, "but we agreed that I'm a fast learner and I can pick that stuff up."

In fact, Beale stated, his only expertise was that he had "done a lot of negotiating in my law profession." Still, once hired, he worked on amendments to the Clean Air Act and was good enough at his job that the team he worked with got the EPA's Gold Medal for Exceptional Service for its efforts.

By 1989, Beale was a full-time federal employee with the title of senior policy analyst within the Office of Policy Analysis and Review—even though he was ambivalent about his latest career choice. "I didn't plan on staying. I was living in a rented apartment with rented furniture, and I was looking around, making plans to go to other places and to do other things," he said in a deposition.

To keep him from accepting any of several alleged private sector job offers, in 1991 the EPA gave Beale a "retention bonus," as recommended by his supervisor, Brenner. That bonus amounted to 25 percent of his annual salary, and Beale received it for not just one but three years. There would be additional retention bonuses in future years. We discovered those were paid mainly because nobody remembered to stop paying them.

Over the next decade, Beale was a noteworthy employee. During the October 1, 2013, hearing, Brenner testified that "John had established a track record that made him one of the most highly regarded members of the EPA. He developed many strong relationships at EPA, on the Hill . . . and he became a frequent and well-respected participant in clean air strategy meetings at the White House." According to the *Washingtonian*, "EPA employee Lydia Wegman, who worked with Beale until the mid-'90s, would later laud his charisma and gift for mastering complex issues: 'He could . . . explain them clearly and forcefully to others both within and outside EPA, and marshal persuasive arguments in support.'" Beale himself claimed that "I had earned the trust and the respect of people at all levels in the organization, career employees, political appointees, Republican and Democratic, and had been very successful in accomplishing a lot." All this good work and goodwill resulted in three EPA gold medals, two commendations—the Lee M. Thomas Excellence in Management Award and the Fitzhugh Green Award for Outstanding Contributions to International Environmental Protection—and a 1999 promotion with the new title of Senior Advisor to the Assistant Administrator.

And then, in 2000, things started to change. Suddenly this unremarkable-looking middle-aged man did not turn up at his EPA office on certain Wednesdays. His electronic calendar explained the absence as "D.O. Oversight," supposedly meaning that he would be at the Directorate of Operations, or CIA headquarters, in Langley, Virginia. That happened 9 days in 2000, 15 days in 2001, 22 days in 2002, 14 days in 2003, 18 days in 2004, 25 days in 2005, 3 days in 2006, and 1 day in 2007—which adds up to 107 days over 7 years. In 2008, he took off nearly all of the six months from June to December, only occasionally showing his face in the office.

Did anybody notice? Did they miss him? Or ask why he was absent? "Because of my personality and because of the fact that I traveled a lot for international work and because a lot of times . . . when you're working with Hill staff and members, people from agencies don't have that show up on their calendars because they're confidential sessions," Beale stated in his deposition.

Now, you'd think that when somebody doesn't show up for work, the logical question might be . . . where are they? Wrong! Not the federal government. The logical conclusion among some federal employees? They are in the CIA!

There were rumors in the office that Beale was a spy for the CIA. "People would joke about it and I would deny it and I would laugh it off or I would say something like, well, if I told you anything I'd have to kill you. I mean, it was a—it was a joke."

But then it wasn't a joke. Beale told at least one boss, Assistant Administrator Jeff Holmstead, that he did work for the CIA. He testified that he said, "'Jeff, I've had this experience working before, working for the CIA, and they've asked me if I would on a limited basis help out with reviewing operations.'" Holmstead not only bought the story, he never asked for any documentation of Beale's secondary employment. Moreover, when Beale pitched a special research project in 2005 that would allow him to work from home, Holmstead and another supervisor gave it the green light. This project, which Beale said he described as a way to "kind of modify the capitalist system to achieve the very goals we wanted," would have three phases and would take "100 percent" of Beale's time for at least three years. He would work at home "in order to not be constantly interrupted with colleagues having questions and wanting to talk to me about things."

Beale had a series of supervisors over the years. The CIA excuse satisfied most of them when they bothered to ask why he wasn't at work. He lied to Assistant Administrator Bob Meyer in 2008, telling him "I was going to be working on a special process for the agency on executive protection. . . . I fabricated that story," he told us in his deposition.

And then we have a May 2010 email to his boss Gina McCarthy—yes, the same Gina McCarthy now criticizing President Trump and EPA administrator Scott Pruitt—that read, "Gina, contrary to what I believed when we spoke last Tuesday, I do have to travel out of the country next week. Events last weekend have made this trip necessary. I expect to be back in about ten days." McCarthy, this savvy custodian of America's air and water, replied, "Thanks John. Stay safe."

Another time he fabricated an overseas trip, saying "I had to make a fast trip to London last night. Still here but heading home in a few hours. Will be back around 2100 tonight and will be in the office tomorrow. Sorry for this diversion." At other times he claimed to be in Pakistan or "in the tank," referring to being at the CIA.

So, what was Beale actually doing? He admitted to us that he would be at home in Virginia where he would "read, bicycle [or] work on the house." A logical question would be, didn't his family inquire why he was spending so much time at home? According to his deposition, he began lying to his wife, Nancy Kete, as early as 1994, telling her that he worked for the CIA. She had been an EPA employee as well but moved to New York to take on a position at the Rockefeller Foundation in 2012. As of his appearance before our committee, the two were still married.

Faking his attendance record was one thing, but that's not the only way Beale defrauded the EPA. This paragraph comes from the sentencing memorandum filed in the case of the *United States of America v. John C. Beale*: "In addition to his theft of time—

amounting to approximately two-and-a-half years between 2000 and 2012—Mr. Beale has acknowledged that he billed the EPA $57,000 for five trips to California over a two-year period that were unnecessary to the performance of his Agency work and that he should have paid for himself because the travel was made for personal reasons."

The funds (spent on hotels and first-class plane tickets that Beale claimed he needed because he had a back injury) covered John's flights to California, where he visited family members in Bakersfield.

There was also the matter of Beale's handicapped parking spot at the EPA. He asserted that he needed one because he had contracted malaria when he was a soldier in Vietnam. The problem? He never fought in Vietnam and he never had malaria. Again, nobody asked him for documentation. But he told us in his deposition that he knew the symptoms of malaria; "having been a medic in the Army, I treated people with malaria." The ultimate cost of that parking spot, by the way, was eight thousand dollars.

McCarthy, Beale's final EPA supervisor, decided to bring the spy-who-never-was in from the cold in December 2010. That's when she called him to terminate his long-term "research project." He told us, "She asked me to come back to be working full-time on kind of the regular duties and reassume management of the international work and the climate work for the air office, and basically said things were so busy that we just can't afford having somebody out there doing an academic project. We need all hands on deck."

Beale complied—at least for a while. Then he decided to retire. The February 4, 2014, memorandum from the Senate Republican staff of the Committee on Environmental and Public Works states, "May 4, 2011, McCarthy approved a draft email to be sent to all OAR[Office of Air and Radiation] staff announcing Beale's imminent retirement from the Agency:

I'd like to express my appreciation to JB for managing OAR's international efforts these past months while we worked through an important period of leadership transition. . . . John will now turn his attention to a few projects where his expertise and experience can continue to add significant value. As you know—John has been a vital part of EPA and the OAR leadership for more years than he cares to remember. He is beginning to look forward to his retirement in the near future—but thankfully has agreed to work on some key efforts in the near term.

Beale wasn't the only EPA retiree at the time. His old friend and mentor Robert Brenner decided to hang it up as well, and the two planned a joint retirement party, along with a third colleague. The shindig, which took place on September 22, 2011, was reportedly a lavish affair (charged to Brenner's wife's credit card) on a yacht that cruised the Potomac River with more than one hundred EPA employees and their guests. Among them was McCarthy.

You think the saga ends there? Nope. Beale supposedly retired at that time. But more than a year after the fancy party, he was still getting a paycheck. Not just any paycheck. He was paid his full $206,000 salary, a sum that included that 25 percent retention bonus. In fact, he was making more than McCarthy. Why was he still getting paid? He never submitted his retirement paperwork.

Ultimately, Beale was caught and prosecuted and pleaded guilty to felony theft of government property. He repaid the whole $886,186 before going to jail in 2013 and liquidated his retirement accounts to pay an additional $507,207 in penalties. He was sentenced to thirty-two months in federal prison and released on June 1, 2016. In his deposition, he told us that he perpetrated the fraud out of greed, and that he always thought there was a "good likelihood" that he would eventually get caught. He added, "it kind of becomes like an addiction. I'm not saying it's an addiction but it's similar properties, and I think I made up my mind several times to stop but it never succeeded."

If his abuse of the system weren't so egregious, it might have been funny. Jon Stewart found humor in the situation in a segment about our hearings on *The Daily Show*. In his inimitable sarcastic style, Stewart commented, "This is what's so wonderful about this story. This man is a liar and boring as [bleep]. . . . That's what is so fascinating here. It's an amazing fraud perpetrated by a guy so he could do things we only do when we've run out of other things to do! I will commit fraud and then, eh, read a book, maybe ride a bike. There's nothing on television. Then I'll make some juice from concentrate. I don't know. So again, it's brazen criminal, ordinary life. He's the secret life of Walter [bleep] Mitty."

I know of no better story for expressing the lack of accountability you see all over the Deep State. You see, Beale did not make it all the way to retirement because he was incredibly good at this. He made it because the Deep State takes care of its own. Beale was punished only because his behavior went past the egregious to the just plain nuts.

But wait. There's more, as they say. Remember Beale's good friend and recruiter Robert Brenner? Apparently Brenner had been totally in the dark about Beale's nefarious activities. We found out at the hearing that Beale just happened to have spent the past two weeks in Brenner's guest room after having minor throat surgery. When I heard that during the hearing, I could only sputter, shake my head, and say, "Mr. Chairman, this is just an unbelievable story. I yield back."

Stewart used that clip of me on the show and shouted as if in disbelief, "This guy Beale just broke the House Oversight Committee! Look at Chaffetz's face! He can't take it anymore. He's like, sorry, guys, I gotta tap out. Lucky for him, somebody else is ready to tap in, baby." Then Stewart played a clip of my colleague Elijah Cummings, Democrat of Maryland, who asked Brenner with an incredulous look, "Beale is staying in your guesthouse?" Brenner answered, "Mr. Cummings, Mr. Beale needed a place to live in the area. . . ."

As for Beale's supervisor, Gina McCarthy, on July 18, 2013, the Senate confirmed her as the thirteenth administrator of the Environmental Protection Agency, a post she held until Donald Trump took office on January 20, 2017. I had several more encounters with her over the years. Read on.

Sexual Harassment and Porn at the EPA

While John Beale may have been one of the most outrageous cases of abuse at the EPA, sadly it was not the only one. During my time on the Committee on Oversight and Government Reform, we held at least four hearings looking into mismanagement and incompetence at the EPA.

One of the agency's middle managers, a man named Peter Jutro, was the subject of our gathering on April 30, 2015. I started the hearing by saying, "Mr. Jutro was the acting associate administrator for the EPA Office of Homeland Security. He also happens to be a serial sexual harasser."

Patrick Sullivan, the assistant inspector general for investigations in the EPA's Office of Inspector General (OIG), testified before us that day. In his statement he wrote, "In August 2013, we received an allegation that Jutro, the Acting Associate Administrator for the EPA Office of Homeland Security[,] engaged in a series of interactions involving a 21-year-old female intern from the Smithsonian Institution . . . who reported him to a supervisor and indicated that she was uncomfortable and scared."

Jutro's side of the story, according to Sullivan, was that

his initial contact with the intern took place on or around July 16, 2014. Jutro confirmed that he gave the intern his business card, so that she could contact him regarding career questions. He stated that, after several failed attempts to meet, they finally met for lunch at a restaurant on July 30, 2014. Jutro stated, during lunch, that he invited the intern to his office, so they could talk where it was less noisy. Once in his office, Jutro confirmed he asked the intern what turned her on and what excited her.

He explained to the OIG that he asked these questions from a career standpoint. Jutro admitted he took a photograph of the intern's face and toes.

Contrary to what the intern said, Jutro denied brushing up against her, attempting to kiss her, or grabbing her buttocks but conceded that he might have put his hand on the small of her back as they walked through security. He admitted to wiping something off the intern's face during lunch but he did not feel it was inappropriate.

However, we found that wasn't the only time Jutro engaged in what was, even in the best light, inappropriate behavior. Continued Sullivan, "From 2004 through July 2014, Jutro engaged in unwelcome conduct with 16 additional females, which included touching, hugging, kissing, photographing, and making double entendre comments with sexual connotations."

Sixteen more women. At least thirteen of them made formal complaints to their bosses or appropriate senior-level officials at the EPA.

What happened to Jutro? Was he reprimanded? Fired? No. He got promoted. Finally, when he was confronted with the allegations and asked to submit to questioning, he avoided due process by retiring—with his full government pension, no less. These days he is a private consultant. His website, where he blogs about subjects like the future of bees, understanding the farm-to-table movement, and the tree of life, touts his tenure with the EPA yet makes no mention of his ignominious departure from the agency.

Perhaps that's because he didn't see anything wrong with his behavior. *Mother Jones* included an email from Jutro in a September 2, 2016, article, in which he denied wrongdoing. "It is true that I have hugged many people, both men and women, and have done so since childhood. My parents were German Jewish refugees who detested the coldness of their former country in the 1930s and strongly encouraged this warmer behavior in me. I also learned to sometimes kiss a person on the cheek or head as a greeting or farewell. In no case was there ever a sexual component to this. I recognize in retrospect that my behavior might have made someone uncomfortable and I feel bad and embarrassed about that, but it was never my intent. There may be actual sexual harassment at EPA, but I was not a part of it."

More infuriating is the fact that the EPA management did nothing about this man. When confronted about EPA's lack of action, supervisors either tried to justify the lack of oversight or offered no plausible explanation for it.

It would be one thing if Jutro were the only serial sexual harasser at the EPA who committed his crimes and paid no penalty for them. Unfortunately, there are others. Only four months later, on July 29, 2015, we found ourselves holding yet another hearing about sexual harassment, cover-ups, and employer retaliation at the EPA.

This case centered on an EPA scientist named Paul Bertram. Like Jutro, Bertram was an older man (sixty-two) who forced his attentions on a younger woman. This time she was a twenty-four-year-old research fellow. As I explained in the hearing, "[T]his perpetrator, Mr. Paul Bertram, inappropriately hugged her, rubbed her back, grabbed her, rubbed her hands, touched her knees, kissed her, made suggestive comments, and engaged in unsolicited physical and verbal contact."

This happened countless times over a period of years. The consequence? They moved the sixty-two-year-old's cubicle.

This wasn't the scientist's only offense. Ross Tuttle served in human resources in the EPA's Region 5 office, where Bertram worked. It was his job to investigate the complaint, he told us. "I obtained the names of more than a dozen other female interns that had worked in this same organization going back to the year 2000. . . . From their statements, I learned that not only had this employee been systematically sexually harassing female interns (going back to at least the year 2000), but an email I received from a current employee and male colleague of the perpetrator stated that the perpetrator had been 'disciplined' by his university for this same kind of behavior during his Ph.D. program."

Tuttle and several other Region 5 employees graphically testified about how the district's senior management retaliated against them for blowing the whistle on Bertram. When they reported the harassment, or tried to do their jobs, these employees were shouted at, told to "stand down," instructed to violate the law, and called "insubordinate" for refusing to do so. Worse, they were shuffled into dead-end positions and given meaningless tasks that gave them no chance for advancement. Behind their backs, they said, management called them "troublemakers" and "activists." Their careers were derailed, they were reassigned, or they were demoted.

I questioned EPA administrator Gina McCarthy, John Beale's former boss, about this when it was her turn to testify at the hearing. These are excerpts from our exchange.

CHAFFETZ: Do you believe the three witnesses that were here . . . do you believe they were retaliated against?

McCARTHY: No, I do not.

CHAFFETZ: We heard nearly two hours of testimony, and you believe that their statements are false.

McCARTHY: No, I did not indicate that.

CHAFFETZ: Well, they said that they were each retaliated against, and you said that is not the case.

McCARTHY: I indicated that what I look at is the entire facts around the case. And, clearly, we had confusion in how we investigated it, but they were part of a large team that actually recommended removal of that employee, and they are no longer in federal employ. . . . There was nothing to retaliate, and retaliation will not be tolerated.

It went on and on like this. I talked about all the other cases and the fact that none of the cases was referred for criminal prosecution even though the conduct was against the law.

McCarthy had nothing but excuses and bureaucratic-speak. I finally had this exchange:

CHAFFETZ: It is against the law. It is against your own policies and procedures. What these people testified to is they had to step up, go to the mat, and say—offering a reprimand is not sufficient.

McCARTHY: The employee was removed, sir, not reprimanded. He was removed from service.

CHAFFETZ: The problem is they had ten victims to get to that point. Now, I grant it, you were not the administrator the entire time. I understand that. But this predator, the quote we heard, this was a predator who was fed a steady diet of interns. The first time it happened he should've been fired, and he should've probably been referred to the authorities for criminal prosecution. It happened ten times, and you never did that.

McCARTHY: I am aware that eleven years ago there was an issue raised. And it was handled appropriately, is my understanding—

CHAFFETZ: Appropriately? He got a promotion. He continued to work there.

McCARTHY: He was carefully watched. The very minute we had any indication of impropriety, which was the recent issue, we took prompt action.

CHAFFETZ: You moved his cubicle four spaces away. You think that is appropriate?

What do you say to the mother and father who sent their twenty-four-year-old to the EPA, she is starting her career, and she is harassed? Look at her statement. And you did the appropriate thing by moving her four cubicles away?

McCARTHY: Human Resources recommended the same thing as every manager, which was to actually proceed to removal.

CHAFFETZ: That is not what initially happened. It was in his record that they had had ten complaints, ten sexual harassment complaints, against this gentleman, and he was allowed to continue to be there. And, as we heard testimony, a predator who was fed a steady diet of interns.

I said at the time, and I still believe it's true, "One of the most toxic environments we have is at the EPA. The mission of the EPA is to protect the environment, protect the people. The problem is the EPA doesn't protect its own employees." As we have seen, it doesn't police its employees, either.

In our April 20, 2015, hearing, besides looking into Peter Jutro's misdeeds, we also discussed three EPA employees who watched pornography at work.

One of them, fifty-four-year-old Thomas Manning, from Lafayette, Indiana, was an EPA information technology (IT) specialist. He used one of the agency's laptops to surf the Internet for pornography. By his admission, he collected more than fifty thousand images and videos. Before returning the laptop to the EPA, he tried to erase his illicit activity. However, some evidence of that activity was uncovered when the computer was reassigned to another employee, who reported it to his supervisor. Besides the agency laptop, Manning kept a personal hard drive locked in a drawer in his office with tens of thousands of pornographic images.

Manning was caught, tried, convicted, and sentenced to thirty months in federal prison. He was one of the few who faced the consequences of their actions. At our hearing, we were told of an unnamed employee who downloaded seven thousand pornographic files onto an EPA server and twenty thousand more onto his laptop. When questioned by a special agent for the OIG, he admitted that he watched between two and six hours of sexually explicit material daily while at work. Unbelievably, he told the agent he didn't think he was doing anything wrong because all his assigned work was done. Even more unbelievably, Inspector Patrick Sullivan informed us, "During the period he viewed the pornography, he received several performance awards, including awards up to $2,000 and a time off award."

This employee was put on paid administrative leave in May 2013. It took nearly two years—until March 2015—for the U.S. attorney's office to decide they weren't going to prosecute him. The following month he retired.

So far, we've seen how the Deep State wastes money and does their best to keep outsiders from seeing what they're up to. Here we see what happens when their problems are uncovered: paid administrative leave, retirement with full pension, or nothing at all.

Sullivan went on to discuss another unnamed employee whose hardcore viewing habit was exposed by a child who was in the office during Bring Your Daughters and Sons to Work Day and happened to see what was on his laptop. This guy claimed he only watched pornography at the office for one to four hours a day.

"About 3,500 pornographic images were recovered from the employee's laptop and external media," Sullivan testified, and approximately 30 to 40 percent of the data on his external media devices was devoted to indecent material. Like his smut-viewing colleague, this employee went on paid administrative leave and was not prosecuted by the U.S. attorney's office.

I noted at the hearing, "These people were being paid roughly in the neighborhood of about $120,000 a year. One of these employees finally retired after almost a year of paid leave. The other employee is still collecting his government salary and he too has been on paid administrative leave for almost a year. American taxpayers continue to pay this person. If you sit watching hours of porn on your government computer, fire them. Fire them. . . . This pattern of paid administrative leave followed by retirement with full benefits is totally and wholly unacceptable. It rewards bad behavior and leaves taxpayers footing the bill."

My colleague on the Oversight Committee Elijah Cummings was equally repulsed by what we found out about the EPA. He took Acting Deputy Administrator Stanley Meiburg to task.

CUMMINGS: If I were watching this, Mr. Meiburg, just watching C-SPAN, I would be disgusted. We are better than this. We are so, so, so much better. Just think . . . the sexual harassment issues. Are you married, Mr. Meiburg?

MEIBURG: Yes, sir.

CUMMINGS: The idea that your wife would come to work, after doing all the things she has to do to get ready in the morning and take care of her family, then she has to come and be harassed . . . Man, you would go crazy. . . . The idea that you have these folks who stay in the employment of our EPA, after having done these things, I just cannot get past . . . Something is missing and we are better than this. We are so much better. If you cannot do the job, you need to let somebody else get in there and do it because a lot of people are depending on government functioning properly. They just want to come to work, do their job, give them their blood, sweat, and tears and then go home, but then their morale gets destroyed when they see people coming back to work, they will get a little tap on the hand, come on back, welcome, watch some more porn. Give me a break, this is crazy. We are better than this.

The EPA, with a culture that takes a laissez-faire attitude toward sexual harassment, fraud, and other misconduct, a management that fails to act, and a lack of consequences for wrongdoing, is the perfect embodiment of the problem with the Deep State. Despite our best efforts at oversight and reforming this agency, very little has changed. In November 2016, yet another EPA employee in Region 5—where sexual harasser Paul Bertram worked—was indicted for having pornography on agency devices. Floyd O'Hara, sixty-two, allegedly stashed his illicit materials on eight EPA computers and servers and when caught, tried to destroy the evidence.

I was asked about this latest incident and commented, "The problems at EPA Region Five persist and worsen with each account. It is deplorable for an agency to have sunk to such toxic levels. Leadership from the very top is needed to restore the integrity of this office."

The only good news is that this employee was not put on paid administrative leave. He was arrested and prosecuted.

In a war with few victories, this was a small but happy one.

CHAPTER 4

The Deep State Fights Back

One of my biggest problems—or I should say one of the biggest problems the Deep State had with me—was that I wouldn't just accept their attempts to evade congressional oversight. I took my role as the chair of the House Oversight Committee very seriously. The American people are footing a tremendous bill for the federal budget, for millions of employees, contractors, and consultants. I understand the federal bureaucracy has become a fact of life. However, it doesn't need to be out of control. It shouldn't—and must not, if our democracy is to survive—be corrupt and dishonest. I still believed it was my job every day to ensure that the bureaucracy was running as fairly and efficiently as possible.

Obviously, that oversight role often placed me and other members of Congress in an adversarial role with the agencies and federal employees we investigated. But that natural tension is exactly what the Founders wisely designed—a system of checks and balances between the branches of government.

After all, the Swamp likes to keep us from physically inspecting their work, fails to punish malefactors, and wastes money at an improbable rate.

Might exposing that involve disagreement and discord? Probably.

But that is Congress's job. That disagreement must be transparent and open and fair.

That brings me to the topic of the U.S. Secret Service. First, a little history of this agency, which currently employs 6,500 people and has an annual budget of $1.6 billion.

The Secret Sex Service

This is how the Secret Service describes itself: "The United States Secret Service, one of the nation's oldest federal investigative law enforcement agencies, was founded in 1865 as a branch of the U.S. Treasury Department. It was originally created to combat the counterfeiting of U.S. currency—a serious problem at the time. In fact, following the Civil War, it was estimated that one-third to one-half of the currency in circulation was counterfeit."

So, while most people think of the Secret Service as the agency protecting the presidents and their families, it has always done a lot more. It is charged with far more authority than many people realize, and arguably its tasks are broadening every day. In 1996 the Secret Service was given the authority to investigate fictitious financial instruments of almost any sort; in 1998 the Telemarketing Fraud Prevention Act was passed, allowing for convictions associated with fraud in telemarketing. In 2003 the Secret Service was transferred from the Treasury Department to the new Department of Homeland Security, specifically to protect the critical financial infrastructure of the United States.

By 2008, over a five-year period, the Secret Service had made 29,000 arrests for counterfeiting, cybercrimes, and other financial misdeeds. Their conviction rate was 98 percent. By 2010 the service announced the creation of its second overseas Electronic Crimes Task Force (ECTF), a network of public-private partnerships dedicated to fighting high-tech, computer-based crimes. There are ECTF offices in the United Kingdom and in Rome.

In other words, the Secret Service is much more than a group of presidential bodyguards. The trust and reliance that the nation places in it cannot be overstated. The soundness of our nation's currency, our global infrastructure—both are dependent on the protection of the Secret Service. Let's call the agency what it is: one of the most powerful intelligence agencies on the planet.

That is what makes the egregious pattern of misconduct by the Secret Service in 2012, 2013, and 2014 so deeply troubling.

In April 2012, President Barack Obama headed to Cartagena, Colombia, for a Summit of the Americas, a gathering of thirty-three national leaders.

As usual the president was preceded a few days earlier by a team of Secret Service agents, most of whom checked into the Hotel Caribe.

What followed can only be described as a night of debauchery, dancing, copious amounts of alcohol, and prostitution. One of the prostitutes, Dania Suarez, actually wrote a book. About twenty prostitutes were involved with at least thirteen Secret Service agents.

There were investigations, confessions, enormous publicity, and both resignations and firings of the agents involved. Apparently this kind of behavior by agents protecting the president and representing the United States was not a one-off. Stories later emerged of earlier parties in China and Romania.

President Barack Obama, asked about all of this on Jimmy Fallon's late-night television show, glibly called the agents "knuckleheads." To me that was a tepid condemnation, hardly proportionate to the egregious misconduct.

Later, in October 2014, it emerged that there had been an actual cover-up at the White House, according to the *Washington Post*.

"New details drawn from government documents and interviews show that senior White House aides were given information at the time suggesting that a prostitute was an overnight guest in the hotel room of a presidential advance-team member—yet that information was never thoroughly investigated or publicly acknowledged," the *Post* reported.

In fact, the investigation into the entire incident was stymied by the White House. "We were directed at the time . . . to delay the report of the investigation until after the 2012 election," David Nieland, the lead investigator on the Colombia case for the DHS inspector general's office, told Senate staffers, according to the *Post*.

Here is another core idea every American needs to understand: the Deep State's interests almost always align with a politician eager to cover it up. When a government employee does something wrong, his superiors look out for him, and the political officeholder above the superior is usually more than happy to see it fall down the memory hole.

The scandals just kept rolling along. In 2013, the management of the Hay-Adams Hotel, just a block from the White House, notified the Secret Service that an agent was attempting to reenter the room of a guest in the hotel, a woman he had met in the bar. She refused to let him in. Inexplicably, hotel security found a bullet in the room.

And then we have the March 4, 2015, incident at the White House. At about 10:25 p.m., a woman in a blue Toyota drove up to the southeast entrance of the White House on Fifteenth Street. She got out of her car with a package wrapped in a shirt and announced it was a bomb. She laid the item on the ground and returned to her car. Agents ran to the car, opened the passenger door, and managed to put the car in park. But the woman managed to put the car in reverse, accelerated, hit both the agents and a barrier, and sped off.

Police secured the area. All the precautions one can imagine were put in place while an inspection team was called in to examine the package on the ground. Yellow tape was set up, a perimeter was established—all the things one would expect when someone has driven onto the White House grounds and left what they said was a bomb.

Not far away that same evening, at the Fado Irish Pub, a retirement party was being held for a government employee. The party, consisting of about thirty to forty people, featured an open bar and was set for 5:30 to 7:30 p.m. Two Secret Service agents, Marc Connolly, a high-ranking agent with twenty-seven years of experience in the Presidential Protection Division, was at the party. Another agent, George Ogilvie, who had been with the service since 1996, also attended.

After the party ended, the pair and two non-agents stayed on at the bar, opening an alcohol tab with a credit card. All four left at 10:45 p.m., with Ogilvie driving his government-issued SUV carrying Connolly back to the White House to retrieve his car, which Connolly had left parked there.

At about 11 p.m., Ogilvie and Connolly approached the White House, showed their passes, then drove around the barriers, through police tape, past a temporary barricade, and into a barrel. They essentially drove through a crime scene, almost hitting the object that could have been a bomb. Supervisors at the scene believed the agents were drunk. Police wanted to arrest the agent. Incredibly, a more senior officer told them to let the agents go, according to the *Washington Post*.

All of this happened just a month after Joseph P. Clancy had been appointed the new director of the Secret Service, ostensibly to clean up the troubled agency.

Is this the way a top U.S. security service should be operating? Just wait. It gets worse.

Of course, it was the job of the House Oversight Committee, along with the Department of Homeland Security's inspector general, to investigate the incident.

The Deep State Comes for Me

At 10 a.m. on March 24, 2015, the committee convened to question Secret Service director Joseph Clancy about the incident, particularly the allegations that the agents breached a crime scene and were under the influence of alcohol.

I began questioning Director Clancy just after the hour.

By 10:18 a.m., a senior Secret Service agent working in the Office of Administration at headquarters logged on and put my name into the MCI Secret Service database. That database has been described as a 1980s-vintage system of records that houses information on a variety of people, including those under investigation, personnel, and past and present job applicants. It can contain personally identifiable information such as birth dates, Social Security numbers, and medical records.

There was no reason in the world to rummage through the database at that time looking for information on me—no legitimate reason, anyway.

Many years ago, I had applied for a job at the Secret Service. I confess I was not entirely serious at the time. I did not get the job, which was for the best. I think history has shown my talents can be better used elsewhere.

But within minutes, this agent found my old application. He called another agent in the Dallas field office, who then also accessed my application. By the end of the day, seven other agents had accessed my file. By the end of the next day, March 25, 2015, another thirteen agents had poked around. All told, the Department of Homeland Security's inspector general, examining computer records, found that forty-five Secret Service employees had accessed my record sixty times. Who were they and where were they? Well, a partial list:

Office of Government Affairs
Dallas Field Office
Phoenix Field Office
Charlotte Field Office
London Resident Office
Washington Field Office
Albany, Georgia, Resident Office
Sacramento Resident Office
Office of Strategic Intelligence and Information
Counter Surveillance Division
William Clinton Protective Division
Indianapolis Field Office
San Francisco Field Office
New Haven Resident Office
Boston Field Office

One agent told the IG that while he was in New York City the next day, guarding the president of Afghanistan, many of the seventy agents present were talking about it. The IG reported that disclosure of my information had likely spread to hundreds of people, both in and out of the Secret Service.

On March 31, assistant Secret Service director Ed Lowery wrote an email to fellow assistant director Faron Paramore. "Some information that he might find embarrassing needs to get out," Lowery wrote. "Just to be fair."

Well, "get it out there" they did. On April 2, 2015, the *Daily*

Beast published the big story. Someone leaked the earth-shattering news of me applying to the Secret Service in 2002 or 2003 and being turned down. Headlined "Exclusive," the story implied I was harboring some decades-long bitterness against the Secret Service rather than doing my job as a congressman.

When the *Daily Beast* reporter Tim Mak called me, he asked whether I harbored "ill will" toward the Secret Service. All I could think of was what I said: "That's pretty funny, no." I told him I might have applied to the FBI, too!

Since this was about me, it's funny. However, a powerful federal intelligence agency violated a citizen's basic rights. Many, many agents were involved. Initially Director Clancy said he wasn't aware of the breach until April 1 . . . but in October the *Washington Post* reported that Clancy told investigators he heard "rumors" about the breach on March 25. He did nothing.

My colleagues on the House Oversight Committee were outraged, as were numerous colleagues on both sides of the aisle. Maryland representative Elijah Cummings, the ranking Democrat, was especially furious.

If federal agency employees can pull these illegal tricks on a congressman, imagine what they might do to a regular citizen?

So what happened to all the agents who violated essential trust in a critical intelligence agency—who broke laws?

On May 26, 2016, Reuters stated that Secretary of Homeland Security Jeh Johnson reported that forty-one Secret Service agents had been punished. How were they punished? Were they charged with a crime? Fired? If you've been paying attention so far, you know there were no such meaningful penalties.

Their "punishments" ranged from a letter of reprimand to suspensions without pay for up to forty-five days. One agent resigned.

CHAPTER 5

The War on Whistleblowers

The Department of Justice is without question a pillar of the Deep State. The situation sadly continues today under President Trump and Attorney General Jeff Sessions—unless Sessions is gone by the time you read this.

Deep State entrenchment in the DOJ is of particular concern because we count on Lady Justice to be blindfolded—objectively dispensing justice according to the law. The Justice Department is, arguably, the first line of defense against abuses both inside and outside the government. The DOJ should be protecting us. And yet it is the federal agency that stands head and shoulders above the rest in enabling the Swamp.

Let's start with the numbers: The DOJ is one of the largest federal agencies, with 116,476 employees who receive a base compensation of $10.24 billion (fiscal 2016). Bonuses added another $29 million. The average compensation with benefits is $115,177. The highest-paid employee? The chief psychiatrist for the Federal Bureau of Prisons, who earned $260,000.

As chairman of the House Oversight Committee, I had a unique power. I could unilaterally subpoena anyone. I didn't need a vote from my committee. I didn't need permission from anyone. I could issue a subpoena to anyone. I could unilaterally haul them before Congress to testify under oath. Just as important was the authority to subpoena anyone in government to produce documents I wanted.

If I thought any federal agency was doing something wrong—anything questionable—and Congress wanted to ask about that action, whether it involved the IRS, the State Department, the Department of Veterans Affairs, even something like airline safety through the actions of the Federal Aviation Administration or the Department of Transportation or the National Transportation Safety Board, I could issue a subpoena.

As citizens, you and I would take a subpoena pretty seriously. Of course, we have to comply. A legal subpoena is not notice of an overdue library book. To most of us, I suspect, it would cause alarm. It certainly would for me. In real life, if you ignore a subpoena, you risk going to jail.

Let's assume I want an investigation into possible wrongdoing. I would issue a subpoena to the Drug Enforcement Administration (DEA) or the FBI, or the ATF, Bureau of Prisons or even Interpol.

I issue my subpoena. Now guess who I must rely on to enforce that subpoena? You got it. The Department of Justice! All of those agencies are within the DOJ. So I would be dependent on the DOJ enforcing a subpoena against itself.

Now, let's say that the agencies didn't want to enforce the subpoena. Let's say one of the thousands of career employees who were hired under a different administration didn't like the politics of the current one. Or they didn't like the idea of siding with Congress over their brethren. The federal agency employees would ignore the subpoena, talk to their career pals over at DOJ, and feel pretty comfortable that the DOJ would never enforce it.

That is the way the DOJ subverted congressional oversight time and again.

How Congress Gave Away the Subpoena Power

It wasn't always this way. The seeds of Congress's attitude toward subpoenas were planted in 1857 but sadly didn't become apparent until 1982.

In 1857, Congress passed a law ruling that anyone found in contempt of Congress was criminally liable. That should have strengthened the check and balance. But then Congress agreed to change the way contempt would be enforced. Instead of holding time-consuming hearings, Congress agreed to certify matters of contempt to the district attorney for the District of Columbia, "whose duty it shall be to bring the matter before the grand jury for their action."

That was the first mistake. The second was failing to rectify the error when the Reagan administration first discovered the utility of it in avoiding prosecution. Prior to that time, congressional threats to certify a complaint to the district attorney generally resulted in cooperation. Between 1975 and 1998, there were ten votes to hold cabinet-level officials in contempt. All resulted in complete or substantial compliance with information gathering. But in 1982, a congressional oversight subcommittee subpoenaed records from the EPA regarding mismanagement of EPA Superfund hazardous waste sites.

President Reagan's then EPA administrator, the late Anne Gorsuch, refused to comply with the subpoena. Gorsuch, whose son Neil currently serves on the United States Supreme Court, claimed executive privilege. Now we know where Eric Holder learned this trick.

In keeping with the 1857 agreement, the Speaker of the House quickly certified the matter for prosecution by the district attorney. Only this time, the district attorney refused to take the case to a grand jury, exposing a massive weakness in what was once a secure check and balance.

Over the next few years, the executive branch's own Office of Legal Counsel affirmed the right of the district attorney to exercise discretion to prosecute. That created a carve-out for executive privilege by directing Congress to file civil suits to enforce subpoenas. Congress quietly went along with this erroneous interpretation of inherent contempt. We've been paying for it ever since.

The Fast and the Foolish

Perhaps no story better captures the sheer intransigence of the DOJ than the epic battle in Congress over what came to be called the Fast and Furious scandal, where the U.S. government allowed two thousand guns to be put in the hands of drug cartels. The tale is complex and detailed, in part because there are so many layers of politics, cover-ups, and schemes to maintain the status quo. The whole awful story is not just about guns, ineptitude, and bureaucracy. It is about President Obama's administration setting up a program that inevitably failed to protect America not just from illegal immigration, but specifically from dangerous criminals and drug smuggling. It is about an attorney general and a Justice Department that hid evidence, lied, and defied Congress. And ultimately, it is about a government that cost a brave border patrol agent his life.

First, a little background on the ATF. The Bureau of Alcohol, Tobacco, Firearms and Explosives—known as the ATF—was formed in 1886 to be something of an über–law enforcement agency that could help and coordinate with local police. Seemingly depending on whatever crime was particularly in fashion among lawbreakers—mob crime, money laundering, alcohol, gun-running, drugs—it bounced around in the jurisdiction of the Treasury Department, the IRS, then was briefly part of the FBI during Prohibition. After September 11, 2001, the ATF was permanently returned to the DOJ.

Back in 2008, an idea was hatched at the senior level of the Phoenix office of the ATF. Mexican drug cartels seemed to be importing guns from the United States into Mexico, and those guns were then used again in crimes in the States. In May 2008, William Newell, special agent in charge of the Phoenix ATF office, told ABC News: "When 90 percent–plus of the firearms recovered from these violent drug cartels are from a U.S. source, we have a responsibility to do everything we can to stem the illegal flow of these firearms to these thugs." Good idea. But like many good ideas, the execution of this one had major flaws. Somebody's bright idea was to allow U.S. gun dealers to sell guns to the cartels and then "follow" the guns to see where they were involved in crimes. It was called "gun-walking"—follow the guns to their ultimate destinations when they were used in crimes. Now, you might wonder how one does that successfully without round-the-clock surveillance. Simple. You don't.

There were several iterations of this flawed policy, but the most dramatic failure was the operation nicknamed Fast and Furious.

For context—and remember, this example is in just one border area in one state—let me give you a brief timeline since 2009 of the kinds of incident reports compiled by the Santa Cruz County, Arizona, sheriff's office, and reported by Leo W. Banks in the *Tucson Weekly*. These reports concern Peck Canyon and surrounding areas, twenty-five miles from the city of Nogales, Arizona:

- November 21, 2009: Men dressed in black fire at eight illegals in Peck Canyon with automatic weapons, wounding David Luna Zapata. He runs through the mountains for an hour before reaching Jim Cuming's house and calling for help.
- November 27, 2009: A hunter in Fresno Canyon, two miles north of Peck Canyon, finds the body of a Hispanic male, shot to death. When Border Patrol backtracked, they spot "four scouts in a cave and attempted to apprehend the subjects." The agents also "spotted three subjects walking on the same route with seven subjects approximately ten minutes behind." The men are believed to be drug *mules*.
- December 27, 2009: A sniper shoots a Border Patrol agent in the ankle in Ramanote Canyon, on the Lowells' ranch, about two miles southwest of their home. Two agents had entered the canyon after spotting a subject "possibly carrying a marijuana bundle."
- June 11, 2010: Two men wearing camouflage and masks fire on seven illegals near Ramanote Wells, on the Lowells' ranch. As the illegals flee, they meet two more masked men, who also fire "multiple rounds at them." Manuel Esquer Gomez, wounded in the arm, says that as he and the other illegals flee, they stumble across a decomposing body with the head and hands missing, "possibly due to animal activity." The body is two hundred yards from the Lowells' home. Pima County's Medical Examiner tells the *Weekly* the dead man is too "heavily skeletonized" to determine a cause of death.

- July 2, 2010: Hilltop gunmen fire at ten illegals, likely in Peck Canyon (based on descriptions). Two aliens say they couldn't see the assailants, but bullets hit the ground around them. One man, running "as fast as he could down the canyon," is hit in the back. The illegals admit entering the U.S. three days earlier to find work in Tucson.
- July 7, 2010: Based on a reliable intelligence source, Immigration and Customs Enforcement warns that a bounty has been placed on Nogales Border Patrol agents. The alert says twenty to twenty-five snipers, possibly from the Beltrán-Leyva Cartel, are headed to Nogales, Sonora, to shoot agents. The alert says snipers would be paid $5000 for each person shot and cautions agents "to remain vigilant, maintain awareness of their surroundings, and utilize body armor and long arms as appropriate."
- August 28, 2010: Two men in camouflage carrying handguns approach two Hispanic males between Peck Canyon and Negro Canyon. The gunmen ask "in Spanish for the marijuana." The men say they don't have marijuana and flee. The incident occurs a mile from the Lowells' home, on the north side of Peck Canyon.
- September 5, 2010: Snipers fire multiple shots, probably with a high-caliber rifle, at Border Patrol agents in Bellota Canyon, north of Peña Blanca Lake. An agent returns fire; no one is hit. Border Patrol agents also took fire inside the Peck Corridor on August 9, 2009, near the town of Ruby, and on June 21, 2010, west of DeConcini port of entry in Nogales.
- September 12, 2010: Three men carrying rifles and wearing bandana masks fire at two illegals in the Walker Canyon area, northeast of Peña Blanca Lake. No one is

hit. One of the illegals says the assailants were in a "very green" area and "suspects they were growing marijuana." The victims entered the U.S. west of the border wall and walked through a large hole in the barbed wire fence, "big enough where ATVs have been passing through." They'd followed the ATV tracks for an hour when attacked.

- September 14, 2010: Seven illegals report being assaulted near Atascosa Lookout, six miles west of Agua Fria Canyon. Jesus Enrique Perez-Mercado says the gunmen stole $200 from him and assaulted him when he balked at giving up his rosary beads. Perez-Mercado says the gunmen spoke English to each other but Spanish to their victims.

- October 21, 2010: A man in a hooded jacket and carrying a *cuerno de chivo*, slang for AK-47, attacks nine illegals—five men and four women—in the Pajarito Mountains south of Peña Blanca Lake. He robs them, kicking some of the men in the stomach. He tells the women to strip and penetrates them with his fingers. He separates out one woman and rapes her, never removing his hood or relinquishing his weapon.

- November 11, 2010: Three men wearing masks rob an illegal at gunpoint on Wise Mesa, in the national forest between Peck Canyon and Agua Fria Canyon. The assailants tell the man to leave the area and not return.

- November 16, 2010: Border Patrolmen on horseback encounter twelve illegals two miles northwest of Peña Blanca Lake. An agent shoots one of the illegals in the stomach after the illegal reportedly threatens him with a rock.

In other words, the area was rightly called a major smugglers' paradise. This was a secret to no one.

The Cover-Up Begins

Border agents were unsupported, overwhelmed, and underarmed to deal with criminals and illegal immigration. Journalist Sharyl Attkisson, then a top reporter for CBS News, identified at least a dozen cases where federal agents allowed guns to "walk" in places like Florida, New Mexico, and Texas, under covert programs called "Too Hot to Handle," "Wide Receiver," and "Castaway." The feds encouraged U.S. gun dealers to sell guns to suspected traffickers to the cartels. The goal was to see where the weapons ended up and then make a big case against the cartels. That never happened, of course.

The Peck Canyon area was under siege, much like many other areas on our border.

On the night of December 14, 2010, after a few days in the area searching for smuggling activity, a U.S. Border Patrol Tactical Unit (BORTAC) encountered a "rip crew"—slang for armed bandits who rob drug smugglers without the bother of having to cross the border—in Peck Canyon. They called out "Border Patrol! Don't move!"

A firefight ensued. Border Patrol agent Brian Terry, a Marine Corps veteran, was hit by a single bullet from an AK-47. The remote and rugged area, which now included armed criminals, was too dangerous for a medevac helicopter. Border Patrol agents carried Brian Terry for more than a mile to a safe place for the chopper to land. It was too late; Brian Terry's spinal cord had been severed, along with an artery to his heart. Shortly after midnight, Terry lost consciousness and died en route to the hospital, according to testimony from fellow agents on the scene. He was forty years old.

Two AK-47-type assault rifles were recovered at the scene of Terry's murder and traced to Operation Fast and Furious. A straw purchaser with known connections to the Mexican drug cartels had purchased the firearms from a shop in Glendale, Arizona, on January 16, 2010. ATF had been aware of the purchaser's connection to a straw purchasing ring since it conducted surveillance of him on November 25, 2009. However, pursuant to the Fast and Furious operational strategy, ATF agents took no actions to disrupt the straw purchase.

At the time of Terry's murder, officials at ATF, the U.S. attorney's office, and the Justice Department knew about the tactics forming the basis of Operation Fast and Furious. In fact, on December 15, 2010, the day after Terry's murder, ATF connected the firearms found at the murder scene to Operation Fast and Furious. But no one told the Terry family. Josephine Terry, Brian's mother, first heard about Fast and Furious from reporters who started calling for her reaction.

In the subsequent weeks and months, the Terry family searched for answers about Brian's murder. On February 8, 2011, Carolyn Terry, Brian's stepmother, emailed Republican senator Charles Grassley of Iowa and asked for help. She wrote:

It's hard to accept that our son was shot and murdered with a gun that was bought in the U.S. We have not had any contact from the Border Patrol or any other agents since returning home on the 22nd of [January]. Our calls are not returned. I truly feel that our son's death is a cover-up and they hope that we will go away. That will not happen. We want to know who allowed the sale of that gun that murdered our son. Any help will [be] appreciated. We are the victims of this case and we want some answers.

Senator Grassley included that email in a letter to Obama's attorney general Eric Holder on February 9, 2011. Documents show

Holder agreed with Grassley that the Terry family deserved answers. According to an email from Gary Grindler, Holder's chief of staff, to Assistant Attorney General Lisa Monaco, Holder was "particularly concerned" about "the assertion that there has been no contact with the victim's family." Monaco replied: "It sounds like [D]ennis' office has been in contact with the family and that there are multiple factions in the family."

However, officials in the U.S. attorney's office seemingly disregarded Holder's message that the Terry family deserved answers. Justice Department officials discussed the Terry family disrespectfully. For example, in a February 23, 2011, email, an official in the office of the U.S. attorney for the District of Arizona referred to the family as "disgruntled."

A disturbing level of contempt for the family was documented in a June 2017 Joint Staff Report issued by the House Oversight and Judiciary Committees. According to the report, emails show DOJ staffers deliberately limited the amount of truthful information to share with the Terry family, and made those determinations based on concerns about media coverage. Days after U.S. attorney Dennis Burke's meeting in Michigan with the Terry family, he and his colleagues were discussing their strategy for communicating with the Terry family. Assistant U.S. Attorney Jesse Figueroa emailed Burke and several others:

> We are making a mistake by attempting to reason with the stepmother and the brother. Lisa and I have been dealing with them since the start of this case. . . . Stepmom is irrational and I firmly believe she and the brother enjoy being in the limelight. Whatever they are told will not change their irrationality and will just cause them to contact the news. If they learned about our hope for a wire I have no doubt

*that would have been on the news also. We should deal only
with the intelligent side of the family.*

Furthermore, the DOJ lawyers denied that the weapons used to
kill Terry were part of Fast and Furious. In June 2011, the Terry
family was still searching for answers.

According to the Justice Department's Office of the Inspector
General, under operations "Fast and Furious," "Too Hot to Handle," and "Wide Receiver," indictments show that the Phoenix
ATF office, over protests from the gun dealers and some ATF
agents involved and without notifying Mexican authorities, fa-
cilitated the sale of more than 2,500 firearms (AK-47 rifles, FN
5.7 mm pistols, and .50-caliber rifles) to traffickers destined for
Mexico. Many of these same guns are being recovered from crime
scenes in Arizona and throughout Mexico, which is artificially
inflating ATF's eTrace statistics of U.S. origin guns seized there.
One gun is alleged to be the weapon used by a Mexican national
to murder Agent Brian Terry on December 14, 2010. ATF and
DOJ denied all allegations. After appearing at a congressional
hearing, three supervisors of Fast and Furious (William G. Mc-
Mahon, Newell, and David Voth) were reported as being transferred and promoted by ATF. The agency denied the transfers
were promotions.

Turning on Their Own

One of the real dangers of confronting the Deep State is retribu-
tion. I experienced it as a congressman in the Secret Service scandal. But the true victims are the honest federal employees who see
wrongdoing, become whistleblowers, and find themselves punished for it.

In June 2011, Vince Cefalu, an ATF special agent for twenty-four years who in December 2010 exposed ATF's Project Gunrunner scandal, was notified of his termination. Two days before the termination, Representative Darrell Issa, Republican of California and at the time chairman of the House Committee on Oversight and Government Reform, sent a letter to the ATF warning officials not to retaliate against whistleblowers. Cefalu's dismissal followed allegations that ATF does just that. ATF spokesman Drew Wade denied that the bureau is retaliating but declined to comment about Cefalu's case.

Enough was enough. This is about a number of issues: border security and immigration, drug smuggling, and the Obama administration's failure to tell the truth.

I have often called my way of learning things "management by walking around." I've always done that, and I did it a lot as a congressman. There is nothing like going to a place, talking to people. You get beyond the debriefing by the so-called experts who tell you only what they want you to hear and believe.

Not that most people in the Deep State are happy to see me.

So, I started making trips to the U.S.–Mexico border. Between April 2011 and June 2015, I made five trips. On some of these I issued no press releases. But I did take photos documenting what I saw. I spent time with landowners who showed me bullet-ridden areas of their property. With local police and border agents, I viewed evidence of crimes, including fake identification documents, stolen goods, and more bullet-ridden items.

On one trip near Naco, Arizona, I saw sixty unaccompanied elementary schoolchildren leaving school on the U.S. side and crossing the border home to Mexico. No border checks, no papers.

And this happens daily.

Several of those trips were arranged for me by various federal agencies. The senior brass would line up, give me nice briefings with a PowerPoint presentation, take us on tours in air-conditioned vehicles. I got the stories of how everything was going well.

If the border was more dangerous in the wake of Fast and Furious, I sure didn't hear about it.

In April 2013, I arranged through friends a private trip to the Yuma area. I took an after-hours unofficial ride along with Border Patrol agents. Nothing in the official briefings even hinted at the things I would see that night with my own eyes.

Being a Border Patrol agent is a grueling, difficult, and dangerous job. They are not well compensated—something I would later pass legislation to help address. But they face danger every single day. They have never been able to let their guard down on the job. But in the wake of an administration reluctant to enforce the law, everything had changed. I came expecting to hear about the consequences of a failed gun-running operation, but I saw the very real consequences of a Deep State colluding with a president to undermine existing law.

We were in separate vehicles just getting ready to leave the station in Yuma for the border when the first call came in.

There were nine people—Romanians, they thought—coming across the border. We were called to apprehend them. My adrenaline was pumping. I was all excited, but these Border Patrol agents were just sitting there—in no hurry at all!

Why weren't they gearing up for the chase?

"Don't we need to hustle?" I said to the agent next to me. "Shouldn't we be doing lights and sirens and getting there as fast as we can?"

They all started laughing. I'm thinking, *You've got to be kidding me. They're on the run, and we need to go after them or we'll miss it.*

"Jason," the agent replied, "they don't run."

"What do you mean they don't run?" I asked, bewildered.

"They want to get caught. They've waited their whole lives to get caught," he said.

Then he explained how it works now. They get caught. We give them water and food—make them comfortable. Then they claim asylum. A few days later, they'll go before a judge, who will tell them to come back in six years for a court date. Meanwhile, an attorney will help them apply for a work permit. Suddenly, they will be legally in the United States with access to education, health care, government benefits, and the ability to compete against Americans for jobs.

I was shocked. Sure enough, the nine Romanians did not run.

Incredulous, I asked the agent why we wait six years to give them a hearing on their asylum claim. I learned we only have three administrative judges in all of Arizona. There had been such an increase in asylum claims that dates for hearings stretched to 2020 (this was in 2013). The Romanians we apprehended could expect to be legally working in the United States within a matter of weeks after being caught crossing the border illegally. No wonder people were coming from around the globe to claim asylum. I learned this was happening by the tens of thousands.

It got worse during the Obama administration. During previous administrations, the Border Patrol agents told me, there had been a general feeling that the sophisticated drug cartels—the ones with superior technology and resources—didn't want trouble. They didn't want to make headlines that might bring more scrutiny, more security, and more problems for them on the border. They recognized that sometimes their people would be caught— and that was the cost of doing business across the border.

But now they recognized that people were being rewarded, not penalized, for crossing the border. They didn't need to be so careful. And they could step up their operations.

Even more worrying was the impact on minor children. As drug cartels learned that an ostensibly soft-hearted Obama administration would send minor children home without penalty, they would purposely traffic kids. They would send young minor children with eighty-pound packs on their backs across a dangerous desert, knowing they would be greeted with food, water, and safe crossing once they reached the border. They knew we wouldn't arrest the kids—we would send them back home.

Anyone could cross the border safely and comfortably, including a terrorist or a criminal.

The U.S. border is vast and unprotected in many places. President Trump's call for a wall—a real wall—is not political rhetoric. Anyone who travels the length of the border can see that.

For example, there is a small part of the river on one section that is very easy to wade across. The water is warm, fairly shallow, and has no predators for them to worry about. And it is even comfortable in places. In fact, someone—I didn't know who—had actually installed a walkway coming out of the water. Not just any walkway; this one was fully ADA-compliant. There were handrails. You could have crossed it in a wheelchair! There was no fence on this part of the border at all.

The Brian Terry murder was not the only instance of guns from Fast and Furious being used in crimes, and not even all of them could be tracked. On February 5, 2011, ICE agent Jaime Zapata, thirty-two, was killed in Mexico by members of the Zeta drug cartel and his partner wounded. The assailants carried AK-47s and handguns.

We even heard testimony at one point from an agent who told Congress, "[T]here was this sense like every other time even with

Miss Giffords' shooting, there was a state of panic like, oh my God, let's hope this is not a weapon from the case." Of course, he was referring to the shooting of Representative Gabrielle "Gabby" Giffords in Tucson in 2011.

While Holder and DHS secretary Janet Napolitano issued a press release on February 16 forming a joint task force to investigate the Zapata murder, saying they had met to discuss it . . . Holder testified under oath that he and Napolitano never discussed Fast and Furious.

The whole thing—the lies, deceptions, and cover-up—was sickening to me.

In March 2018, Kent Terry, the brother of slain agent Brian Terry, tweeted and asked President Trump to fulfill a campaign promise to reopen the Fast and Furious case.

I agree. I tweeted "I met with AG Sessions to get the Fast & Furious documents. He said NO and decided to let it continue to play out in court. Frustrating and disappointing to say the least."

CHAPTER 6

Contempt of Congress

Operation Fast and Furious and its predecessor programs were not simply what liberal outlets such as the *New York Times* repeatedly called "botched gun trafficking" operations. They were not management failures by poorly supervised yahoos at the ATF.

These programs, and the epic effort by the Obama administration to conceal them, to lie about them, to resist openness and truthfulness to the American people, to defy the constitutionally mandated efforts of a duly elected Congress to learn about wrongdoing, were nothing less than an epic effort by the Deep State to defy the will of the Founders.

That may sound like hyperbole. Unfortunately, it is not. I was there.

Inventing Privilege

Attorney General Eric Holder and a slew of Department of Justice witnesses testified for nearly a year about these programs before our committee in Congress. Holder himself testified nine times before Congress. They lied and lied and lied, as was revealed later in a 471-page report by the DOJ's own inspector general. They drafted a letter telling Congress that the ATF was confiscating guns. The exact opposite was happening. They concealed documents. They didn't want to tell Congress why this crazy program was hatched, and was allowed to go on, even when they realized that weapons were not being tracked. They didn't want to tell Brian Terry's family why he had been killed.

President Obama gave the go-ahead for the use of "executive privilege" to protect Holder from having to reveal documents. That was pretty creative and unusual, as executive privilege had only been used to protect the president, not a cabinet member or a federal agency. In fact, many of us, as well as impartial commentators, remarked that the fact that the president tried to extend *his* executive privilege to Holder and the DOJ suggested that the president himself knew about Fast and Furious and the internal communications about the program. Congress wasn't even asking about communications between the DOJ and the White House.

Of course, first Holder tried to assert something called "deliberative privilege," a common-law privilege not recognized by Congress.

As you can imagine, I was angry and frustrated.

"You have a dead U.S. Border Patrol agent, you have thousands of weapons that were knowingly given out by the Department of Justice to the bad guys. We gave them to the criminals. We have nearly three hundred dead in Mexico and a host of questions that the Department of Justice has never, ever answered," I said.

Kimberley Strassel called this tactic Holder's "Privilege Creativity" in her June 26, 2012, column in the *Wall Street Journal*. She reminded readers of Holder's similar efforts in 1998. Back then, Holder was deputy attorney general in the Clinton White

House during prosecutor Kenneth Starr's investigation. Holder tried to create and assert an entirely new privilege called "protective privilege," which would forbid Congress from requiring truth-telling testimony from Secret Service agents protecting the president. Strassel called the proposed privilege "legally nuts." A federal appeals court unanimously agreed and rejected its legality.

On Wednesday, June 20, 2012, the House Oversight Committee, then chaired by Representative Darrell Issa, voted Holder to be in contempt of Congress.

A week later, on Thursday, June 28, 2012, after an eighteen-month investigation, the full Congress voted to hold Holder in contempt, by a vote of 255–67. Some 17 Democrats voted in the majority, and 108 Democrats abstained.

"Claims by the Justice Department that it has fully cooperated with this investigation fall at odds with its conduct; issuing false denials to Congress when senior officials clearly knew about gunwalking, directing witnesses not to answer entire categories of questions, retaliating against whistleblowers and providing 7,600 documents while withholding over 100,000," said Chairman Issa.

This was the first time in U.S. history that an attorney general had been held in contempt by the full Congress.

The Teapot Dome Scandal

Attorney General Eric Holder was not the first cabinet official to be found in contempt of Congress, but he was the first attorney general to earn that distinction.

That first honor belongs to Secretary of the Interior Albert Fall.

This audacious Deep State plot took place in 1921. Before Watergate, the Teapot Dome Scandal was, as reported on history.

com, "the greatest and most sensational scandal in the history of American politics."

The Teapot Dome Scandal of the 1920s shocked Americans by revealing an unprecedented level of greed and corruption within the federal government. The scandal involved ornery oil tycoons, poker-playing politicians, illegal liquor sales, a murder-suicide, a womanizing president and a bagful of bribery cash delivered on the sly. In the end, the scandal would empower the Senate to conduct rigorous investigations into government corruption. It also marked the first time a U.S. cabinet official served jail time for a felony committed while in office.

The "ornery oil tycoons" were Henry F. Sinclair, who owned the Mammoth Oil Company, and Edward L. Doheny, from the Pan American Petroleum and Transport Company. Both were leased the exclusive rights to federal oil reserves. Sinclair got the ones in Teapot Dome, Wyoming. Doheny got the ones in Elk Hills and Buena Vista Hills, California. They obtained the leases from Secretary of the Interior Albert Fall (one of the "poker-playing politicians"). The problem with this deal was that Fall did it in secret. There was no competitive bidding. He also had received a $100,000 "loan" from his old friend Doheny, which was delivered in cash in a satchel (the "bagful of bribery cash") by Doheny's son, Edward Jr. (Ned), and his close friend Hugh Plunkett.

The shady deal appalled both conservationists and other oil dealers who complained so loudly that the Senate looked into the matter. Fall took a "nothing to see here, move along" stance and almost got away with the misdeed. Until the "loan" came to light.

In 1924, Doheny testified before the Senate committee and admitted that he had indeed given $100,000 to Fall. Sinclair said he had given Fall some livestock for his cattle ranch. Both oil barons claimed the gifts were totally unrelated to their leases. Nonetheless, the leases were canceled and Congress asked the president for a special prosecutor to investigate further.

Warren G. Harding was the ("womanizing") president when the leases were originally granted. However, he died of a heart attack on August 2, 1923, in bed, while listening to his wife, Florence, read him a flattering article from the *Saturday Evening Post*.

The investigation was left to his successor, President Calvin Coolidge, to carry out. In the end, which came in 1929, says historian Cherny, "Fall was convicted on charges of accepting a bribe from Doheny—the only guilty verdict in the Teapot Dome case. . . . [The] first Cabinet member convicted of a crime committed while in office, [Fall] was fined $100,000 and sentenced to a year in prison." And the murder-suicide? Writes Cherny, "In 1929, Doheny's son was murdered by Plunkett, who then committed suicide; Plunkett may have feared that he'd be sent to prison for helping to deliver the cash to Fall."

There is certainly precedent for holding cabinet officials accountable for failure to comply with the law. But in the case of Fast and Furious, the cabinet official in question, Eric Holder, presided over the agency that would be required to prosecute him.

Holder's Contempt *for* Congress

Contrary to what many in the liberal media thought, I did not enjoy holding the U.S. attorney general in contempt. A top government official, much less the president of the United States, in essence declaring our Constitution a joke, and our Congress a laughingstock, is a dangerous and sad moment. This was not something to celebrate.

The media often portrays back-and-forth like this as "scoring political points." But when someone working on behalf of the people forgets that the *point* of it all is the policies, not the politics, it's time for them to get out of government.

The chief law enforcement officer of the United States, Attorney General Eric Holder, was then asked by ABC News how he felt about the contempt vote in Congress:

"For me to really be affected by what happened, I'd have to have respect for the people who voted that way. And I didn't, so it didn't have that huge an impact on me," Holder said.

Think about that for a minute. Holder, still at the time the sitting attorney general of the United States, told the entire world, both our friends and foes, that he had no respect for the 255 members of Congress, including the seventeen members of his party, who voted to find him in contempt.

I'm sorry. But the word *treason* comes to mind.

Later on, I'm going to talk about other cases, other hearings, and other ways that Obama-era Swamp creatures subverted due process, ignored subpoenas, and defied Congress. But here I want to address the question: does it matter who is president? With a massive federal bureaucracy and a barely submerged Deep State, can someone like President Trump—or President Anybody, for that matter—really change things?

I've mentioned that on March 3, 2018, Brian Terry's brother called on President Trump to release the documents still held for seven years at the Department of Justice. Kent Terry tweeted to President Trump:

Sir it's been 7 yrs. My family ask you reopen Obama's gun scandal that cost my brother his life. I talk to you back on the campaign trail here in Michigan and you offered to reopen the books into this senseless scandal. thank you. God bless.

Just days later, on March 7, 2018, after years of prodding—and quite frankly after both myself and Representative Trey Gowdy kept talking about the issue and criticized Attorney General Jeff Sessions's inaction, the DOJ made an announcement:

Today, the Department of Justice entered into a conditional settlement agreement with the House Committee on Oversight and Government Reform and will begin to produce additional documents related to Operation Fast and Furious. The conditional settlement agreement, filed in federal court in Washington D.C., would end six years of litigation arising out of the previous administration's refusal to produce documents requested by the Committee.

Sessions added:

The Department of Justice under my watch is committed to transparency and the rule of law. This settlement agreement is an important step to make sure that the public finally receives all the facts related to Operation Fast and Furious.

How many years did it take? And how did the media cover the news, the release of the documents? Silence from the *New York Times*. Silence from CNN. Silence from all the "mainstream" outlets. Reuters did a story. And of course Fox News.

Now, about those other cases I mentioned. When Holder told ABC News that he didn't have any respect for Congress, I wasn't the only one outraged. Chuck Grassley from Iowa, a senior and respected senator, told the *Washington Post* that Holder's attitude "may have also led to people in this Administration thinking they can go after conservatives and conservative groups," a reference to the IRS scandal.

No doubt.

CHAPTER 7

Lying to Congress

For many decades, Charles and David Koch of Koch Industries, one of the largest privately owned companies in the world, had advocated policies that could be best described as libertarian-oriented, free-market economics. They helped found the Cato Institute—a think tank—and a host of other free-market groups, including Americans for Prosperity. It is a pretty safe assumption that they would contribute money to various Tea Party groups, which aligned with them on most policies.

And so it was that a little-noticed, initially anonymous quote emerged from a press briefing August 27, 2010, by Austan Goolsbee, a senior White House official who chaired the president's Economic Recovery Advisory Board. Said the "official":

> *So in this country we have partnerships, we have S corps, we have LLCs, we have a series of entities that do not pay corporate income tax. Some of which are really giant firms, you know Koch Industries is a [sic] multibillion dollar businesses. So that creates a narrower base because we literally got something like 50 percent of the business income in the U.S. is going to businesses that don't pay any corporate income tax.*

Alarming. How would the White House know or even surmise what income tax Koch Industries did or did not pay unless the White House had examined their tax returns? Which, obviously, would have been illegal.

Mark Holden, chief legal counsel for Koch, told the *Weekly Standard*, "We are concerned where this [tax] information would have been obtained from." Representative Devin Nunes issued a statement asking the House Ways and Means Committee to look into whether the White House misused confidential tax information.

Thus began an IRS scandal that was, arguably, the most serious one in the agency's history. And let me be explicit about what constitutes serious: Americans must have a certain degree of faith in their government. Our faith in the rule of law and our government institutions is what separates us from the average banana republic. It is this faith that underlies and supports our democracy, our elections, our economic stability, and our currency.

By 2013, a Gallup poll showed only 27 percent of Americans thought the IRS was doing a "good" job. It was the lowest-ranking federal agency. (The Centers for Disease Control was ranked the highest at 60 percent.) The fact is that the Internal Revenue Service is the federal agency Americans fear more than any other.

The IRS is quite proud of this:

On July 1, 1919, the IRS Commissioner created the Intelligence Unit to investigate widespread allegations of tax fraud. To establish the Intelligence Unit, six United States Post Office Inspectors were transferred to the Bureau of Internal Revenue to become the first special agents in charge of the organization that would one day become Criminal Investigation. They formed the nucleus that built the Intelligence Unit into an elite group of highly trained, dedicated professionals, who are recognized as the finest financial investigators in the world.

The Intelligence Unit quickly became renowned for the financial investigative skill of its special agents. It attained national prominence in the thirties for the conviction of public enemy number one, Al Capone, for income tax evasion, and its role in solving the Lindbergh kidnapping. From these promising beginnings the Intelligence Unit expanded over the intervening decades, investigating tax evasion by ordinary citizens, prominent businesspersons, government officials, and notorious criminals.

In July 1978, the Intelligence Unit changed its name to Criminal Investigation (CI). Over the years CI's statutory jurisdiction expanded to include money laundering and currency violations in addition to its traditional role in investigating tax violations. However, Criminal Investigation's core mission remains unchanged. It continues to fulfill the important role of helping to ensure the integrity and fairness of our nation's tax system.

Since CI's inception in 1919 to the present, the conviction rate for Federal tax prosecutions has never fallen below 90 percent. This is a record of success that is unmatched in Federal law enforcement.

The IRS actively encourages citizens to report anyone they might know and suspect of not reporting income or exaggerating deductions. On their website, the IRS also opines a bit on antitaxation groups and their viewpoints, disparaging them:

Complicated arguments against the American tax system are built by stringing together unrelated ideas plucked from widely conflicting court rulings, dictionary definitions, government regulations and other sources.

They also call out certain groups and their leaders who argue against taxation:

Though the leadership of these movements used different arguments to gain followers, they all share one thing in common; they received substantial sentences in a federal prison for their activities.

Now, I am not opposed to paying my fair share of taxes, or to laws requiring all citizens to do the same. But consider the weight of those words from a federal agency that has the right to potentially put you in prison. That is pretty strong and intimidating language. Elected officials, including Senator Ted Cruz, have called for abolishing the IRS. Should Senator Cruz be sent to prison for that opinion?

The problem with the IRS goes beyond the political misuse of a government agency. As you'll soon see, it was covered up, the perpetrators received paid leave and cushy retirements, the DOJ failed to punish anyone, and Congress was blocked from seeing key evidence. On top of that, though, they added a new twist: lying to Congress with impunity.

Weaponizing the IRS

Nothing should scare us more than the use of agencies like the Internal Revenue Service and Department of Justice to retaliate against those whose political views run counter to the president's. The breaching of the wall between the political and the administrative agendas of federal agencies has far-reaching implications for every single American. I wish I could say we had fixed it. But nothing has been fixed. President Trump could engage in the same activities tomorrow and there would be no law to stop him. The only difference is that perhaps this time the left would recognize the significance of the threat.

The seeds for the Obama administration's blatant misuse of the government's tax collection authority were planted with a landmark January 21, 2010, U.S. Supreme Court ruling. In a 5–4 decision, the Court held that the free speech clause of the First Amendment prohibits the government from restricting independent expenditures for communications, meaning advertising, from nonprofit corporations, for-profit corporations, labor unions, and associations. The *Citizens United* case allowed everyone to contribute money to political campaigns.

To the extreme left, the sky fell. The hyperbole, passion, and intensity that greeted this Supreme Court decision continues today, but in 2010 you would have thought a nuclear bomb had destroyed the integrity of the American election systems.

A year before the Supreme Court decision, a newly inaugurated President Obama was pushing the American Recovery and Reinvestment Act, a package costing close to $1 trillion. CNBC personality Rick Santelli took to the airwaves from the floor of the Chicago Mercantile Exchange to rail against the act as floor traders began to vocally boo the government's policies. Santelli said, "We're thinking of having a Chicago Tea Party in July!"

Most observers would call that the beginning of the modern-day Tea Party, a movement critical of both Republicans and Democrats who for too long rubber-stamped ballooning debt, government overspending, and government intrusion into everyday Americans' lives.

The *Citizens United* decision freed up groups and associations to spend money on elections. So . . . the backstory.

We all remember what happened next. In 2013, the IRS revealed that its staffers had singled out conservative groups with names like "tea party" or "patriots" that were seeking tax-exempt status. They asked these groups about their members and donors and subjected them to scrutiny that was completely different than other groups. The IRS developed a spreadsheet called "BOLO"—"Be On the Look Out"—for applications from Tea Party–supporting groups. The IRS went through several commissioners, one of whom, Steven Miller, was forced to resign after just six months by Treasury secretary Jack Lew after it was revealed he knew about the targeting.

And, of course, our old friends at the Department of Justice opened a criminal investigation into whether IRS employees broke the law when they targeted the groups. (Try and guess where that investigation went!)

Despite the DOJ's reckless failure to hold anyone accountable, the facts are not in dispute. A 2013 report by the Treasury Inspector General for Tax Administration (TIGTA) described in detail the use of "inappropriate criteria" to screen political advocacy groups. Altogether, 296 such groups were subjected to higher scrutiny.

IRS officials told the inspector general that they had used the keywords of "tea party" and "patriots" as shorthand to efficiently manage a deluge of new political advocacy groups, but that explanation was rejected by the inspector general's office.

"Developing and using criteria that focuses on organization names and policy positions instead of the activities does not promote public confidence that tax-exempt laws are being adhered to impartially," said the report, which Inspector General J. Russell George issued.

President Obama, in a fit of faux outrage, called the report's findings "intolerable and inexcusable." He directed Secretary Lew "to hold those responsible for these failures accountable, and to make sure that each of the Inspector General's recommendations are implemented quickly, so that such conduct never happens again." That was the last America heard from him on that topic.

The Good Luck of Lois Lerner

Despite the fact that there were at one point five different investigations of IRS misconduct taking place, Lois Lerner would somehow escape virtually unscathed. On June 3, 2011, the then chairman of the House Committee on Ways and Means, David Camp, Republican of Michigan, sent a letter to the IRS commissioner at the time, Douglas Shulman, stating that "the IRS appears to have selectively targeted certain taxpayers who are engaged in political speech." Just ten days later the IRS's director of exempt organizations, Lois Lerner, told her IT department that her computer was inoperable. What a lucky break for her. The IRS never was able to recover her hard drive.

Soon after she would tell the House Oversight Committee, "I have not done anything wrong."

But she had. Lerner's group was in charge of processing, then approving or disapproving an organization's applications for tax-exempt status—and applications had flooded in between 2010 and 2012, nearly doubling each year. The Department of Justice has since settled with the targeted entities, openly admitting the political targeting, which Lerner had overseen.

In her May 22, 2013, testimony before the House Oversight Committee, Lerner referenced George's TIGTA report in her opening statement and went on to say, "the Department of Justice launched an investigation into the matters described in the Inspector General's report. In addition, members of this Committee have accused me of providing false information when I responded to questions about the IRS processing of applications for tax exemption." Looking up from her written document, she forcefully declared, "I have not done anything wrong." She concluded her remarks saying that "while I would very much like to answer the committee's questions today, I have been advised by my counsel to assert my constitutional right not to testify or answer questions related to the subject matter of this hearing. After very careful consideration I have decided to follow my counsel's advice." That she did, and the following day she was put on paid administrative leave.

Several important events in this scandal unfolded over the next few months after Lerner's testimony and contributed to her good fortunes. The Deep State always protects its own.

The same day Lerner testified, it appeared the noose was beginning to tighten. The IRS's chief technology officer issued a preservation notice, or "non-destruction order," so that no information would be destroyed. On August 2 the first subpoenas were issued for IRS materials. That created a legal obligation to preserve and produce Lerner's emails. In June the use of those BOLO lists was officially suspended. But Lerner's luck was about to improve.

On June 20 the news was made public that, despite having to furlough its employees for several days that year due to budget cuts, the IRS was going to pay out $70 million in employee bonuses. Among those employees was Lerner, who got $42,000, and former acting commissioner Steven Miller, who had to resign due to the targeting scandal. Miller got $100,000. Meanwhile, Congress threatened to slash the IRS budget even more as punishment for the scandal. Then, just before Christmas, the IRS got a new boss on December 23 when John Koskinen was sworn in as commissioner. Koskinen's subsequent actions would prove very helpful to Lerner.

On Super Bowl Sunday, February 2, 2014, while much of the world was focused on the clash between the Seattle Seahawks and Denver Broncos, two significant events in the IRS investigation took place, both of which would be good for Lerner.

The first was Fox News' Bill O'Reilly's ten-minute interview with President Barack Obama. On this date, there were no fewer than those five open investigations into the IRS targeting of conservative nonprofits. Those included:

- House Oversight Committee
- House Ways and Means Committee
- Senate Judiciary Committee
- Senate Finance Committee
- Inspector General (TIGTA)

Despite those five open investigations, the president stated on national television that the IRS scandal had "not even mass corruption—not even a smidgen of corruption." Obama then admitted that while records show that the former IRS commissioner Douglas Shulman had visited the White House more than one hundred times, Obama couldn't recall speaking to him on any of those occasions.

The second event involved a woman named Catherine (Kate) Duval. Kate Duval was an attorney who had been inexplicably reassigned from the Treasury Department to the IRS. Her job was specifically to handle document production to Congress. She was also a counselor to IRS commissioner Koskinen. Oddly enough, Super Bowl Sunday found her in her office, where, she later told the committee, she suddenly noticed there were all these missing documents. Among those documents were Lois Lerner's emails. Those documents were covered by a preservation order and a subpoena.

As of that Super Bowl Sunday, those documents still existed on backup tapes. Duval made no effort to retrieve them. Just a month later, on March 4, 2014, more than 400 backup tapes were destroyed. It's likely that those tapes contained up to 24,000 of Lois Lerner's emails from the critical period in 2011.

I summarized for the committee on May 24, 2016, the extraordinary series of events that led to the mysterious disappearance of Lerner's emails.

On February 2, 2014, Kate Duval realized some of Lerner's emails were missing from what the IRS sent Congress. The next day, Duval told her colleagues at the IRS about the problems she had found: IT, the Office of Chief Counsel, and deputy associate chief counsel, Thomas Kane.

Many of her emails were missing because her hard drive had crashed in 2011. The IRS knew in early February that there was a problem with the emails.

Koskinen testified that he knew in February. This is what he said at the July 23, 2014, hearing before the House Oversight and Government Reform Subcommittee. "What I was advised and knew in February, was that when you look at the emails that had already been provided to the committee in other investigations, and instead of looking by search terms, looked at them by date, it was clear that there were fewer emails in the period through 2011 and subsequently. There was also, I was told, there had been a problem with Ms. Lerner's computer."

What did Koskinen do about it? He had just learned that the most crucial evidence covered by the subpoena was missing. You'd expect him to spring into action.

According to the inspector general, he failed to look in five of the six places Lerner's emails could have existed—the backup tapes, her BlackBerry, the server, the backup server, and the loaner laptop. In fact, the IRS barely looked for the missing emails at all, according to the IG. Lois Lerner herself could not have scripted it better.

In April, Koskinen's agency notified the Treasury Department and the White House that Lerner's emails were missing. Then they told Congress *in June* by burying a couple of sentences in the fifth page of an attachment in a letter to the Senate Finance Committee. Koskinen came up to Congress to explain what he said. Then he lied.

On June 20, 2014, Koskinen appeared before the House Ways and Means Committee and stated, "since the start of this investigation every email has been preserved, nothing has been lost, nothing has been destroyed."

We now know that to have been patently false. When he testified before the Oversight Committee on June 25, 2015, Timothy Camus, the deputy inspector general for investigations, told us that the employees in Martinsburg, West Virginia, who destroyed the tapes said no one had ever attempted to retrieve them. No one had even communicated the existence of a preservation order or a subpoena for them. Unbelievable.

"The destruction that we're talking about required the employees involved to actually pick up tapes and place them into a machine, turn the machine on to magnetically destroy and obliterate the data," stated Camus. "Our investigation has shown that the two employees who physically put those tapes in that machine are lower-grade employees. . . . They did not stop erasing media . . . until June of 2014. When interviewed, those employees said our job is to put these pieces of plastic into that machine and magnetically obliterate them. We had no idea there was any kind of preservation [order]."

So what ultimately happened to these key players in the IRS scandal? Nothing.

On October 23, 2015, the Department of Justice closed its investigation into the IRS scandal, recommending that no criminal charges be filed. Assistant Attorney General Peter Kadzik's letter to the House Judiciary Committee said, "The IRS mishandled the processing of tax-exempt applications in a manner that disproportionately impacted applicants affiliated with the Tea Party and similar groups, leaving the appearance that the IRS's conduct was motivated by political, discriminatory, corrupt or other inappropriate motive. However, ineffective management is not a crime. The Department of Justice's exhaustive probe revealed no evidence that would support a criminal prosecution. What occurred is disquieting and may necessitate corrective action—but it does not warrant criminal prosecution."

"Disquieting?" Even though the House of Representatives passed a contempt of Congress resolution against Lois Lerner for her refusal to testify, citing the fact that she had waived her Fifth Amendment privilege against self-incrimination by making an opening statement, the Department of Justice didn't do anything about it. On March 31, 2015, U.S. attorney Ronald Machen sent a letter to Speaker of the House John Boehner, Republican of Ohio. Machen wrote, "We respectfully inform you that we will . . . not bring the Congressional contempt citation before a grand jury or take any other action to prosecute Ms. Lerner for her refusal to testify on March 5, 2014." Lerner retired from the IRS on September 23, 2013, receiving her generous taxpayer-funded pension.

Kate Duval went on to work at the State Department, where she was entrusted with document production to the select committee on Benghazi. That couldn't have been an accident. She is now in private practice.

What is Congress to do when the Deep State chooses to protect its own rather than enforce the law? I know what I would have liked to do.

The House Jail

Congress does not throw people in jail. Anymore. But technically, in limited cases and under limited conditions, they can. And they have.

Back in 1795, a man named Robert Randall tried to bribe three congressmen to support a twenty-million-acre land grant with an offer to set aside portions of the land for them. The House ordered the sergeant-at-arms to take Randall and an associate into custody while a committee was appointed to decide how to proceed.

They didn't wait for the executive branch to prosecute. Instead, Randall was tried before the House and found guilty of contempt by a vote of 78–17. He served one week before being released.

In 1800, a Philadelphia newspaper published reports that the Senate found to be "false, defamatory, scandalous, and malicious, tending to defame the Senate of the United States." Publisher William Duane was ordered to attend the bar of the House. He refused. Although the Senate ordered the sergeant-at-arms to detain him, the attempt was unsuccessful. But Duane was later convicted under the Sedition Act by the executive branch, with then–vice president Thomas Jefferson presiding over the affair.

And everybody is apoplectic that President Trump tweets that the media is unfair?

Both houses of Congress believed they had the power to enforce cooperation with Congress. That notion was upheld a generation later by a Supreme Court decision in 1821.

That case involved another bribery attempt in the House that resulted in a charge of contempt. In this case, the accused brought a lawsuit alleging assault and battery by the sergeant-at-arms as well as false imprisonment by the House. The Supreme Court sided with the House.

In the intervening years, this Supreme Court decision has not been challenged. I'm not suggesting Congress just jail people indiscriminately, which would be pretty heavy-handed. But a stronger enforcement mechanism to induce compliance with congressional subpoenas is necessary.

Power to Impeach

Although Congress has initiated impeachment proceedings at least sixty times in our nation's history, only once has a cabinet official been impeached. That happened in 1876. If I had had my way, it would have happened again in 2017.

In the case of John Koskinen, he misled and lied to Congress. When he had information that he knew was false, he never did correct it. I sent a letter to President Obama in July 2015 asking that Koskinen be removed from his position as IRS commissioner but never heard back. I felt very strongly (as did Representatives Jim Jordan, Mark Meadows, Ron DeSantis, Paul Gosar, Gary Palmer, Buddy Carter, and Jody Hice) that we should get rid of him. But how does Congress dismiss somebody? There is a provision in the Constitution that is rarely exercised. The Constitution provides for the expulsion of presidents and judges. We commonly think of impeachment as the tool to do that. But a careful reading of the Constitution also allows for the dismissal of civil officers.

I got online, did my research, and found a wonderful essay on what the Constitution says about advice and consent, written by a constitutional scholar named John McGinnis in 2005 for the Heritage Foundation. The piece discusses the Senate's role in confirming presidential appointments. The president can nominate someone, but nobody gets that job or appointment without the consent of the Senate. The Senate doesn't have to give any reasons for its decisions. They can say they don't like the way he brushes his teeth, they don't care for the way he combs his hair. There is no provision or limitation. It gives the Senate a coequal voice.

The idea is not new. Impeachment authority against executive branch officers has been used before.

William Belknap, the secretary of war under President Ulysses S. Grant, was impeached after a House oversight committee uncovered evidence of blatant corruption during his eight years as a cabinet secretary.

Impeachment was threatened in a subsequent incident foreshadowing today's conflicts. Congress wielded impeachment power in an 1879 dispute with the State Department. In this case Congress took an executive branch officer into custody for failure to cooperate with an oversight investigation.

To gather evidence to support allegations against the State Department's minister to China, the newly created House Committee on Expenditures in the State Department demanded to see records of fees received at the Shanghai consulate during the tenure of George F. Seward. Like another State Department leader in our time, Seward had not submitted those records to the department but had instead kept them in his possession. The committee subpoenaed Seward to appear and produce the records.

Seward did appear before Congress, but his attorney argued the committee had no authority to demand the records. Seward tried to plead the Fifth but the committee argued the plea was invalid because their inquiry was not a criminal proceeding. In response, the committee proposed and the House voted to instruct the sergeant-at-arms to take custody of Seward and bring him before the bar of the House. Seward still refused to cooperate, providing only a written statement that the proceeding was an attempt to impeach him and thus he was not required to witness against himself. That dispute went to the House Judiciary Committee, which ultimately sided with Seward. No impeachment vote was actually held.

One thing is clear from our history: this country has often struggled with scandal. That is not new. What is new is the scope and complexity of our federal agencies, entrenched Deep State interests, and the risks we face. The world was a less complex place in the eighteenth and nineteenth centuries. The capacity of the Deep State to destroy a president, thwart Congress, and disregard the rule of law has never been more severe or held more potential for disaster.

We are Americans. We have to fix it. But how?

Before one can get to strategies and policies and programs, you have to have the people in place to enact those policies. You have to have people willing and courageous enough to stand up to the Deep State, to unsettle business as usual.

Call it, perhaps, the courage to disrupt. It doesn't always come in the package or the person you expect, or even hope for.

The prevailing theory on advice and consent is that it includes authority to dismiss someone if they aren't living up to the duties and obligations of their job. You have a coequal voice to appoint and an equal opportunity to dismiss. But it's a very, very high bar. It's essentially the political death penalty. And I thought Koskinen was worthy of it.

On October 27, 2015, I introduced legislation to begin impeachment proceedings against Koskinen. The resolution, H.J. Res. 494, was referred to the House Judiciary Committee. The core of us in the Committee on Oversight and Government Reform supported it. But quite frankly, I was a little disappointed that a broader group of Republicans didn't support us. Peter Roskam, Republican of Illinois on the House Ways and Means Committee, actually argued that Koskinen was our best foil. He personified what was wrong with the IRS.

Unfortunately, too many Ways and Means Committee members concurred. They liked having him there because they thought he was a weak commissioner. I thought that was a lame approach. But it's one of the reasons the impeachment didn't move forward.

When Donald Trump was made president, he had a chance to send Koskinen packing. He didn't. Ultimately, Koskinen was allowed to serve out his full appointment until November 2017. It was a travesty.

And then, on October 25, 2017, the Justice Department announced that it had settled with conservative groups that had sued over the IRS scrutiny. The settlement was reportedly $3.5 million for more than four hundred Tea Party clients. The IRS, in one of the settlement documents, admitted wrongly treating the plaintiffs and offered a "sincere apology."

For her part, Lois Lerner, who, remember, had retired with her fat bonus, in December 2017 asked a federal judge to seal her testimony that she gave in depositions in connection with the lawsuits. Ms. Lerner complained that revealing her testimony would subject her to harassment. The *Wall Street Journal* wrote, "American taxpayers who will fork out $3.5 million for Ms. Lerner's actions have a right to hear how she justified what she did at the IRS."

John Koskinen eventually retired in November 2017, proud that he had prevailed through all the scandals.

"Survival is its own reward," he told the *New York Times*.

It makes me sick. Essentially, we were right, but there would be no consequences for those involved. The Internal Revenue Service turns people's lives upside down. The IRS inhibited conservative Americans' ability to act on their First Amendment rights. The place was and is such a mess that even the February 2016 computer breach that allowed the theft of 700,000 Americans' Social Security numbers was greeted with a shrug. Oh no, wait. The IRS has offered taxpayers identity theft protection for a year. Free!

Then, at the end of the day, the IRS routinely shrugs its shoulders and admits wrongdoing. I'm not aware of a single person who was demoted or who lost their job or pension in connection with the scandal. How was it so wrong that taxpayers had to pay a settlement, yet nothing happened to anyone? This problem is still not fixed. This could happen again.

"Survival is its own reward." Does any sentence capture the essence of the Deep State better than that?

CHAPTER 8

Face-to-Face with the Deep State

I f not for an extraordinary series of events involving a constituent of mine in Utah, the truth about Benghazi might never have come to light. The world may have accepted a false narrative, the witnesses may have been intimidated into silence by their Deep State bosses, Congress may never have sought emails from the State Department, and Hillary Clinton's duplicity may never have been uncovered.

On the morning of September 12, 2012, I saw news reports of an attack in Benghazi, Libya, the previous night. I had never heard of the city of Benghazi nor could I point to it on a map. But even in a world desensitized to the ongoing violence in the Middle East, a terror attack on the anniversary of 9/11 is bound to stand out. Fairly quickly there were horrifying scenes broadcast of our consulate in flames and people running around with masks and guns in what appeared to be a coordinated terrorist attack.

We soon learned that our ambassador, J. Christopher Stevens, along with security officers Tyrone S. Woods, Glen Doherty, and Sean Smith, had been hunted down and killed that night. For a few days, the details were sketchy. Inexplicably, U.S. ambassador to the United Nations Susan Rice went on national television telling America, and the world, that this attack was absolutely *not* a planned terrorist attack. According to Rice, it happened because of a spontaneous reaction to an anti-Islam YouTube video. The explanation didn't seem congruent with the violent images on our TV screens. For me, and for many Americans, the story didn't add up. It seemed odd at best.

There's something very important to remember about the attack on Benghazi, and far more important, about the explanations offered immediately afterward from President Obama and Secretary of State Hillary Clinton: this country was in the homestretch of President Obama's reelection campaign. Election Day was just weeks away. Although the Benghazi attacks themselves would not measurably impact Obama in that 2012 race, the political calculations made by his team and his Deep State allies during that period would have a profound effect on Obama's successor as the Democratic presidential nominee four years later.

An Unexpected Whistleblower

On September 20, just twelve days after the attacks, I made a decision that would ultimately uncover the tip of a massive Deep State iceberg—one that to this day still has not fully surfaced. On that day, I had no inkling of the chain of events that decision would trigger or just how pivotal it would become.

House Oversight Committee chairman Darrell Issa had tapped me to chair the committee's Subcommittee on National Security, so decisions about whether and how to investigate the Benghazi attacks fell to me. Suspicious about the explanations the Obama administration was proffering, I worked with subcommittee staff director Tom Alexander to draft a letter to Secretary Clinton.

Referred to as a preservation letter, it instructed the secretary of state to produce *all* information related to the terrorist attacks in

Benghazi. It very specifically asked for "all responsive documents in [her] possession, custody, or control, whether held by [her] or [her] past or present agents, employees, and representatives acting on [her] behalf." The letter went on to define "document" to mean any written matter of any nature, including electronic mail (email).

The letter itself goes on for several pages, and I signed it as the chairman of the subcommittee. I had asked Chairman Issa to sign the letter with me, as I thought it would be more powerful if it had the chairman's signature on it, but he declined to do so, and I still don't know why. But perhaps his refusal signals just how insignificant the questions seemed at the time.

While my instincts immediately told me that we—Congress and the American people—were not getting the full story of what happened in Benghazi that night, I never expected what happened next.

At that time my congressional office routinely received roughly ten thousand emails or letters a month. Like most congressional offices, we relied heavily on college interns to sort them out and make sure we got a timely response back to our constituents. It was a massive effort and a hard duty for young interns without much experience.

Shortly after the attacks, an email buried among the thousands of others we receive caught the attention of one of our interns. She had been on the job for less than three weeks. She was brave enough to suggest this particular email should be immediately reviewed by the congressman personally. I'm certainly glad she was astute enough to figure this out and share it with me, because it changed everything.

The email was from a lieutenant colonel in the Utah National Guard. He had previously reached out to other congressional offices to share his story but had had no luck getting an audience with them or even their staff.

He had served for months in Libya as the seniormost military officer and wanted to share his perspective on the Benghazi attack. As the lead American military liaison, he was deeply involved in consulting on security challenges at our facilities in Libya. He completed his assignment in August 2012, just weeks before the attack. According to his letter, he wanted to shed light on the reality of what was happening there and what was not being done.

Because of the sensitivity of the information he needed to share, we could not just chat over the phone or meet in my office. We had to arrange to meet in a secure facility where classified information could be divulged.

The earliest we could get together in a secure facility in Utah was on a Sunday. I communicated with the adjutant general in the Utah National Guard, General Jefferson Burton, who agreed to let us use the SCIF (Sensitive Compartmented Information Facility) at Camp Williams.

I met Lieutenant Colonel Andrew Wood in the SCIF that Sunday morning around nine o'clock. I drove the eight miles from my home over to Camp Williams. I had skipped church to meet with him and he had been excused from his duties so we could meet. We were joined by two other National Guard personnel who sat in on the nearly two-hour meeting.

I don't know what I expected from that meeting. But I left it feeling deeply concerned and worried.

There was a lot we weren't being told. Lieutenant Colonel Wood described and documented repeated terrorist attempts to breach the compound. Other countries had already left due to the danger; we were the last flag flying. Vulnerabilities had been identified and reported. But the State Department refused to provide even the minimum level of security standards required by the Inman Report—a 1985 security standard developed after our U.S. embassy in Beirut was attacked in 1983. Instead, Wood indicated that bureaucrats in the State Department waived requirement after requirement, putting everyone at risk. Both he and Regional Security Officer (RSO) Eric Nordstrom knew this was an unacceptable risk.

I cannot disclose all that Lieutenant Colonel Wood told me, but you'd better believe I lost sleep over what I heard. He knew personally each of the men who had been killed. He grieved their loss. He had worked closely with Ambassador Stevens. Although leaving Libya had not been his decision, he felt tremendous guilt about not being there.

Given the differences between what was being reported in the media and what I heard from Lieutenant Colonel Wood, I knew there was only one way to get to the truth.

I told my wife and Chairman Issa that I needed to go to Libya as soon as possible. I needed to talk to people on the ground in real time and get a sense in person of what we were dealing with at the embassy. I had already learned that the consulate in Benghazi had not been properly fortified yet housed numerous Americans in a dangerous and unstable environment. We also had personnel at our facility in Tripoli, which shared the same vulnerabilities. Even a small, unsophisticated attack could easily penetrate the compound. How could that be?

Screwing Up the Narrative

It has always been my nature to act quickly once I make a decision. That quality hasn't always been congruent with the way government works. But it served me well in this case because I do not believe the bureaucrats were expecting me to show up in Libya just weeks after the attack. I hadn't given them time to get their story straight.

It took a few days after my meeting with Lieutenant Colonel Wood for Oversight Committee staff to make the necessary travel arrangements. With such short notice and an election just weeks away, no one on the minority side of the committee accepted my invitation to join me in Tripoli, Libya. That wouldn't stop them from later crying foul to the media when asked why there were no Democrats on that trip. Classic.

I drove to the airport without fully knowing or understanding where I was going or how I was going to get there. I carry a backpack everywhere I go and keep it fully loaded with my electronic equipment, my medicine, and my paperwork—such as my diplomatic passport. I traveled light.

I was told to get on a Delta flight in Salt Lake City on October 4, 2012, and fly to a European city where I would be met as I got off the plane. I didn't know who would be meeting me, nor was I certain what I should be looking for. They told me not to be worried, as it would be evident. It was.

I then changed planes to board an aircraft to Stuttgart, Germany, where I met up with a member of our committee staff. We drove to our American military base to meet with General Carter Ham. As the four-star general in charge of African Command, often referred to as AFRICOM, General Ham would have been personally involved in the deliberations over whether to send help to Benghazi once the attacks were under way. His story would carry great weight—if he was allowed to tell it.

The military divides up the world in different commands, and AFRICOM is a fairly new one. It handles everything in Africa, except Egypt. It has little to no assets and relies heavily on European Command and Central Command to provide the military aircraft and other assets they need to do their job.

We breezed through several checkpoints and were escorted up to General Ham's personal office. It was a rather large office with big maps and a large, long conference table. The room had the obligatory flags and the pomp and circumstance you would expect. The idea was to get an overview and then first thing in the morning fly with him to Tripoli.

But there was a surprise waiting for me in General Ham's office. Somebody was waiting there who was neither invited nor expected. He was a tall young blond man dressed in a suit. He was standing in the room with the general, a notepad and pen in hand. We were not introduced. I had to ask who he was. The general said he was an attorney sent by the State Department. He was assigned to be present in all our discussions.

I later learned his name was Jeremy Freeman, a thirty-three-year-old who had interned at the United Nations. I would later find out he specialized in Freedom of Information Act (FOIA) requests. This is not surprising. When a federal agency hires a lawyer to specialize in FOIA requests, they aren't there to facilitate transparency. It's a scandal in and of itself that the government is allowed to hire smart, talented people to keep the rest of the government from finding out what they're up to.

This man was not here to help me get to the truth. He was sent by the Deep State to ensure I didn't screw up their narrative. His presence would also intimidate witnesses from being too candid.

I felt blindsided. I had not asked for a State Department minder, nor had I been told to expect one. I immediately recognized that his presence would inhibit my ability to get the truth—which is exactly what he was sent there to do.

In Freeman's presence, the general said he was fine with the arrangement, but he seemed a little agitated. I certainly was. I thought General Ham appeared resigned to having Freeman join us. Frankly, he told me, *my* presence was more problematic than the State Department's. Interesting. People had lost their lives. Why didn't they want the public to know what really happened?

I was accompanied by an Oversight Committee staffer with deep overseas government experience who questioned Freeman's security clearance level. He had sufficient clearance for the meeting with General Ham, so we reluctantly continued with Freeman listening to every word.

General Ham's account probably wasn't the story the State Department wanted told. In that initial meeting in Germany, I asked Ham why the U.S. military appeared not to respond to news that Benghazi was being attacked and overrun. He told me then, as he did later the next day on more than one occasion, the United States had proximity and capability, but the military was not directed to go. It didn't make sense then, and it still doesn't make sense today.

After the attack, the FBI hadn't shown up in Benghazi for eighteen days, but the media walked the grounds of our facility days after the attack. The fact remains that no American personnel outside of Libya were ordered to go into Benghazi to help save the more than thirty people who were under siege. Not at the onset and not during the event. Never.

Obama secretary of defense Leon Panetta gave the world a similar story a few weeks later. He would say, "[The] basic principle is that you don't deploy forces into harm's way without knowing what's going on; without having some real-time information about what's taking place. As a result of not having that kind of information, the commander who was on the ground in that area, Gen. Ham, Gen. Dempsey and I felt very strongly that we could not put forces at risk in that situation."

A few months later, in February 2013, after an overwhelming backlash from the Special Forces community, Panetta would change the story yet again. He testified to the Senate Armed Services Committee that intervention was not possible due to "time, distance, the lack of an adequate warning." It seems Panetta's initial, unscripted analysis more closely reflected General Ham's conversation with me than did Panetta's later narrative. Ham would later change his own story—something I heard firsthand.

Clearly, they knew my presence in Libya so soon after the attacks would give them less flexibility to revise history later. That's why Freeman was there.

When I later learned that this Jeremy Freeman specialized in Freedom of Information Act and congressional requests, I understood that his job at the State Department was to suppress the release of information to the public, the media, and Congress. His job was to use all the legal tools possible to make sure the Department of State and its senior personnel were only viewed in the most favorable light.

It is disgusting, but every department and agency I am aware of employs staff to serve this purpose. From my vantage point, Freeman's mission was to spy on a congressional inquiry, suppress information, and intimidate State Department personnel on the ground.

After our initial hour and a half with the general, my staff person and I exited and went to a local restaurant for some really bad German stew.

The next day, I was scheduled to visit a U.S. embassy team whose ambassador had been murdered and who had lost three more of their own to terrorists. It was a sobering duty. In an embassy the size of Libya, everybody knew everybody. It was a complex with limited mobility. Tripoli was considered a dangerous assignment that precluded personnel from bringing their families along. As a result, this team was a tight-knit group. The loss hit hard.

They deserved for the truth to be told. But I wasn't sure they would be allowed to tell it.

The people at Embassy Tripoli knew precisely who Freeman was and who sent him. It was already obvious at that point that Secretary Clinton and her senior staff had their version of the truth and needed this congressional inquiry to go away.

With only weeks to go until President Obama's reelection, he was desperately trying to sell the narrative that he had won the war on terror. Osama bin Laden was dead and Al Qaeda was on the run! The political fallout from the truth—that four Americans had just been murdered in a coordinated terror attack—was obvious. The truth would also presumably hamper Secretary Hillary Clinton's long-term political ambitions.

In retrospect, I believe the truth would ultimately have been far less damaging to the Obama and Clinton campaigns than the lies. The outcome of Obama's 2012 reelection was, unfortunately, not close. Did any votes hinge on whether or not Al Qaeda was on the run? Still, the Obama administration chose to send Susan Rice out to every news network to lie about a video that had nothing to do with the attack. Even worse, Hillary Clinton reportedly looked the families of the victims in the eye days after the attacks and told them the government would arrest the filmmaker who was responsible for the deaths of their sons. She knew at the time that terrorism, not a filmmaker, was the cause of the attack—a fact that came to light during the 2016 presidential race. Her email the night of the attack to her daughter, Chelsea, revealed she had immediately attributed Benghazi to a terrorist attack.

The Obama administration and its Deep State allies covered

this up out of a reflexive—perhaps pathological—need to whitewash and conceal. The Benghazi tragedy, like Clinton's clumsy use of a private email server, revealed a woman committed to keeping secrets—and a Deep State apparatus all too willing to help.

With our State Department minder in tow, we visited the embassy. I have traveled around the world, visiting nearly one hundred countries in my business, personal, and congressional capacity, and this was my first time as a member of Congress that the State Department had blatantly placed a person in my way. His presence was cumbersome and intimidating. No doubt he was emboldened by the notion that President Obama would win reelection and that Secretary Clinton was going to be the next president of the United States four years later! What did he have to lose?

I had been a constant agitator and visible opponent to the administration. That was my job per the Constitution. Freeman was there to make sure I didn't dig too deep.

Later, the State Department would say everyone was fine with Freeman joining the discussions. Based on my observations that was not true. Absolutely not.

During the day of our trip, we tried to make small talk to learn more about his background, why he was here, and what he was doing. He was exceptionally closed-lipped and shared no details. It didn't take long to realize he was an impediment to learning the truth. His only purpose was essentially to spy on me. During the day I did all I could to evade him and find the truth.

Hidden Witnesses

Despite my success in getting information during my visit to Libya, the State Department continued to go to great lengths to conceal the truth. To be successful in promoting their false narratives, they had to hide the witnesses. They were surprisingly successful in doing so.

Shortly after returning from Libya, I had tried to meet with the heroes who survived that night. But even for me, getting their names proved difficult. The State Department was nonresponsive to our requests for a listing of who had been there that evening. Subsequent transcripts provided in response to House and Senate committee inquiries carefully redacted the names of survivors. Six months after the attack, the public still had not heard from a single witness! CBS News reported that the State Department had been unresponsive to multiple FOIA requests for names and other information. We later learned that witnesses had been required to sign nondisclosure agreements. Having been badly wounded, many would not see service again and would be forced to depend on the federal government to care for their families. CIA director John Brennan denied that there was any effort to discourage witnesses from coming forward. But the implicit pressure to stay quiet was real.

In particular, the contractors were very difficult to find. I learned of one hero who, many months after the attack, was still at Walter Reed National Military Medical Center. I wanted to visit him. But it took me months to even figure out what his name was. I was eventually able to get on the phone with his father and express my thanks and appreciation. I learned his name and that he was still at Walter Reed. His injuries had been severe. I tried to visit. But when I reached the hospital, they swore they had no patient by that name. I later learned his name had been changed to prevent people like me from tracking him down.

I don't know who decided to hide David Ubben—if I had known, I would have hauled that person before our committee to answer for it. But there was no way to tell. I never did end up talking to Ubben. He chose not to join his fellow survivors in writing about his inside account for the book *13 Hours*. But he finally told his story to Catherine Herridge from Fox News in 2013. He later testified against Benghazi attack mastermind Ahmed Abu Khattala in October 2017. His story was heroic and, like the stories of the other heroes that night, deserved to be told.

As a State Department Diplomatic Security Service agent at the time, David Ubben went to great lengths to recover the body of Foreign Service officer Sean Smith. In the second wave of the attack, Ubben was on the roof with Tyrone Woods and Glen Doherty—the former Navy SEALs who lost their lives defending the compound.

More Congressional Oversight Is Needed

There is a game the State Department plays with the House and Senate. Any member of Congress can travel to see how the executive branch is spending taxpayers' money. It is an important part of oversight. As I have said, I believe strongly in "management by walking around." With more than 320 departments and agencies together employing more than two million people, hundreds of thousands of contractors, and dishing out more than $4 trillion each year, there is a lot for the 535 members of the House and Senate to visit, see, and question.

The biggest portion of the discretionary budget goes to the Department of Defense at more than $600 billion per year. There are billions of additional dollars spent on foreign aid, State Department activities, etc.

More members of Congress should be traveling and reviewing how this money is spent. "Sunshine is the best disinfectant" is a mantra to live by, and when members show up asking questions, it is quite revealing what they can learn. However, the Deep State has no desire to be open, transparent, or subject to questioning. Congressional engagement scares the living daylights out of them. Being held accountable feels too new for them.

When members of Congress travel outside the United States, it is the State Department that makes travel arrangements. While the trip is funded by taxpayers, it comes from the State Department budget. They have extraordinary control and insight as to what members see along the way. It takes experience as a member to insist and dictate where we go and what we get to see.

Waiting until the last moment, the State Department will often claim there has been a security threat, and therefore we cannot travel to a planned location. It took a while, but I came to realize that while this was sometimes true, it was often done to shield what was happening. The real purpose of highlighting the potential security threat was to make sure members of Congress didn't see something. It isn't just the State Department that does this but also the Department of Defense and others. I recognize the DOD is in a very difficult position, but that's exactly what representatives and senators should be seeing . . . the difficulties! If we are putting our men and women in harm's way, then members should understand the realities and experience them. Firsthand experience will affect their votes, appropriations, and perspectives.

Often, I would have a prearranged trip with a beautiful, professional agenda. I would let them begin the program but after twenty minutes or so it was time for questions and deviation from the agenda. Yes, I want to see and experience what they recommend, but what I really want to see is what they don't want me to see. Sometimes these detours turn up nothing, but I can't begin to even list the wide array of things we were able to unveil by asking unexpected questions and going places they said I could not go.

If there was a federal employee there, then I expected to be able to go there. My favorite question if it got testy was "What is it you think Congress should not see?" It is an impossible question to try to answer and those who did went down in flames. If they really thought they could challenge me I would lift my phone and say, "Hold on, I want to get this on video. Please tell me your name, title, and what you just told me. I am going to want a real-time record of this for the hearing we are going to have in Washington." It never got to that point. By then they were sufficiently worried and took down their guard.

That tactic worked overseas but does not work, unfortunately, at the main offices in Washington, D.C. The State Department, for instance, is so big and massive I have no idea even where to go to pull the documents I want to see. That is all done via congressional letters and subpoenas.

CHAPTER 9

They Think We Can't Handle Truth

was up early dealing with jet lag and anxious about our trip to Libya. We were picked up from our hotel and brought to the tarmac on the military base.

We boarded a white Gulfstream. The general and I sat on the right side of the plane in the front compartment, knee to knee. He was wearing his full military uniform with the accompanying jacket. I was wearing something much more casual, the equivalent of khaki pants and a long-sleeve shirt and vest. My staff person and Jeremy Freeman the minder sat in the back. The flight attendant, from the military, brought the general his coffee and I stuck with a bottle of water.

We took off from Stuttgart on a beautiful sunny day and headed out over the Alps. Within a fairly short amount of time, we were off the coast of Libya. The flight itself is less than two hours—highlighting the proximity Tripoli had to Europe. The fact is that Libya is not at the end of the world. It was a couple hours' flight from where we had major military assets. And yet . . .

In a short amount of time we were cleared to land in Tripoli, after flying over the coast and then the capital city.

I remembered thinking, as we were landing, that as a young man in 1986 driving from Utah to Arizona, I listened to coverage of President Ronald Reagan's decision to bomb Tripoli in an effort to kill Muammar Qaddafi, or at least get him in line. Qaddafi was funding and participating in terrorism around the world and Qaddafi's antics were a direct threat to the United States of America. In my Honda CRX, I could listen late at night to KNX Radio out of Los Angeles and then a Phoenix radio station for hours when the radio announcer was sharing the details of this attack. It still bothers me to this day that France would not allow our military to fly over France. France made our jets fly around the country on their way to bomb Libya.

Listening to the radio account of the attack and the subsequent pictures from Libya had painted a picture in my mind. It was a land far away, with camels, sand, tents, dictators, and oil making them all cash rich.

Now here I was flying with the four-star general over a country that had been ravaged by war and smothered in sand. The ocean water was a brilliant, beautiful blue, and on the edge of the beaches were numerous oil facilities and homes with a smattering of multi-story buildings. In the distance, the sand stretched as far as the eye could see. Obviously, this was a poor, third-world country.

We landed smoothly on a warm, hot, sunlit day without a cloud in the sky. With the dry desert breeze, I felt like I was back in Yuma, Arizona. The plane taxied near the main airport terminal, where there were a lot of people waiting for us. There was a convoy of fortified Chevy Suburbans and our site security team, including more than a dozen armed guards contracted by the government to protect the general and me. These were Americans who had previous military experience. They knew how to wield a gun and had been highly trained. My eyes and ears were on high alert as I was continually aware of the danger and also doing my best to memorize the details of everything I was seeing.

The local Libyans were also there to welcome us. The door of the plane went down and General Ham went first to greet his Libyan counterpart. They knew each other from previous interactions and were joyfully hugging, as they had known each other for years.

General Ham oversaw Odyssey Dawn, the military offensive the United States carried out in conjunction with our allies to displace Qaddafi. No Americans died in the offensive, which we are all grateful for, but the way we did it still bothers me.

At the time, Libya was not a "clear and present danger" to the United States of America. Under those circumstances, a president, in this case President Obama, should have gone to Congress seeking a declaration of war. It didn't happen.

Nevertheless, as I stood up to step out of the plane, I kept thinking to myself, *Please don't let there be a sniper, please don't let there be a sniper.* I exited the plane and shook the hands of the local Libyans, then the lead of our protection team hurried me into the Suburban in the second row.

I have been in a lot of convoys but nothing like this. Leaving the airport in Tripoli I found there is no rhyme or reason to its parking. You could barely scoot out between cars. In fact, we were pushing other cars out of the way with our front fender in order to progress. Once we got out of the airport and onto the main highway, I was subjected to the wildest ride I've ever experienced. At high rates of speed, we were zigzagging all over the highway, pushing other cars out of the way that were not compliant in yielding the proper space. And I mean literally knocking cars aside.

At one point we took an exit, flipped a U-turn, went back around, and then whipped through some local neighborhood at an amazing speed. It was all necessary, unfortunately, and done to protect all of us. We were being followed. I noticed people in cars behind us trying to record our movement with video cameras. I had no idea who they were or why they were following us.

The farther we drove the more it seemed like we were going the wrong way. We were getting into a more and more rural setting. As we traveled down a dirt road, there was the initial checkpoint for the embassy. The United States had lost its embassy during our bombing campaign. Back then it was located downtown, in proximity to local government buildings. Now the embassy was in a rural setting. The United States had rented a compound of homes in between Tripoli and the airport.

Despite popular belief, the outsides of embassies are guarded by locals and not United States Marines. In the wake of the 2012 Benghazi attack, this policy changed for certain embassies, but not across the board.

One of the more disturbing sights that day was a Libyan guard handling his weapon with sloppy movements as we approached the embassy. His obvious lack of training was not just embarrassing, but dangerous. They waved us through and we drove in to be greeted by the local embassy staff. After exchanging pleasantries, we sat in the equivalent of a large living room, where I got to speak with a good portion of the embassy staff.

Just weeks ago, their friends had been killed. I've had lots of meetings with government employees through the years, but I'd never experienced anything like this. Serving in a post away from their families, having gone through their horrifying attack, they had not been allowed to go home. I sat there for more than an hour and listened to their stories and shared our concern and love from all of us in the United States. Among them was Deputy Chief of Mission Gregory Hicks, an experienced diplomat who was in charge now that Ambassador Stevens had been murdered. It was an emotional time for Hicks and everyone in the room. Nevertheless, they bravely shared their stories and perspectives. It was hard for them to understand how America perceived what they had gone through. They wanted to be home with their families and loved ones, but the State Department was making them stay in place. It was an emotional meeting, and I came to deeply understand the grief and difficulty they were going through. After this meeting, they gave me a tour of the facility.

I had pushed to spend the night at the embassy. I was fighting back on the notion that it wasn't safe. After I toured the grounds, I couldn't believe we were letting anybody stay there.

The embassy had low walls and lacked the fortifications typically found at embassies around the world. Later we would come to learn there had been untold dozens—if not hundreds—of exemptions granted by the State Department to the normal security protocols. This would become the focus of a key hearing with the under secretary for management, Ambassador Patrick Kennedy, and Deputy Secretary of State Charlene Lamb.

Jeremy Freeman, whom I now referred to as the State Department spy, was in tow and within earshot every step of the way. Yet when we met with the Marines, things were different.

I was invited to join a Marine by climbing a ladder to the highest point of the embassy. The Marines had surrounded it with sandbags and a large-caliber gun. They could view 360 degrees and watch for pending attacks. Atop this location, it was just me and the seniormost Marine.

He was frustrated that he and his men were put in this vulnerable situation. The compound walls were not nearly high enough, the barbed wire was insufficient, the cameras and motion sensors were incomplete, and there were trees in the way obstructing their view of potential incoming attacks. Even in Tripoli, they had very few strategic positions to engage an enemy.

To the south of the compound, he pointed out a home that was adjacent to the embassy. On the embassy side of the wall there was a large pile of who-knows-what that was about as tall as the wall itself. He told me that on the day they arrived, there was a ladder propped up outside against the embassy wall. Each day the family at the adjacent home would climb the ladder and dump their garbage into the embassy grounds. I could imagine the pungent odor that would emanate from that garbage heap on a hot summer day in the desert.

The Marine told me the story of how he sent some Marines over there with their weapons to explain that if anything came on our side of the wall, he would have no other options but to shoot them. It truly is amazing our embassy personnel were so politically correct that, before these Marines arrived, they would allow locals to dump their garbage on our embassy grounds. This pile had grown to nearly six feet high and was a total embarrassment.

The embassy did not have sufficient medical supplies, either. It was ill prepared to deal with a rocket attack or large numbers of people attempting to come over the wall. There was not nearly enough secure space in case of an attack. The entire place was a major vulnerability, and this was the new embassy. Marines were exceptionally frustrated and concerned about the vulnerability. The more I learned, the angrier I became.

On the night of the Benghazi attack, a Marine Fleet Antiterrorism Security Team (FAST) was mobilized to make their way not to Benghazi but to Tripoli. As the Select Committee on Benghazi concluded, there was not a single human asset outside of Libya that was ordered to go to Benghazi and help save Americans under attack.

The Marine FAST Team . . . wasn't.

They did not adequately meet the goals set forward to mobilize in a timely way to secure a vulnerable embassy. One of the reasons they were delayed was a dispute between the State Department and the Department of Defense. As the Marines were preparing to deploy, word came from State that they were not to wear their standard military gear. They were instructed to move into their civilian clothes.

What? The State Department was concerned with clothes? Or did the department not want it to appear that Marines, American military personnel, were being deployed to protect a consulate under attack?

This came as a great frustration to the military, who have military uniforms for a reason. Not only are they needed to tell who's who, but they're needed to carry their gear, communications equipment, and other supplies. Yet in a vulnerable situation, where hostilities were likely to break out, the State Department was taking the lead in deciding what our military should be wearing. This stunning decision became a major source of frustration for those who had to go to Tripoli that night.

As few know, after the fight in Benghazi had started, the Tripoli embassy was evacuated. Later the embassy personnel would return to the building I was now visiting. It wasn't safe then, and it wasn't safe while I was there.

As I toured the compound, it became increasingly evident that the bureaucracy in Washington, D.C., had made decisions in the name of expediency that put our men and women in an unacceptably vulnerable position. They knew it. I knew it. With a very real possibility the embassy in Tripoli was going to come under attack, these people would be sitting ducks. There wasn't much our personnel could do to defend themselves, even with Marines. The physical barriers protecting that embassy were poor and inadequate in relation to the potential threat.

Instead of taking notes about the security vulnerabilities or showing concern for corners his agency had so obviously cut, Jeremy the spy from the State Department was tenaciously documenting what I saw and what was said to me.

After lunch I was given an opportunity to have a very highly classified briefing from the seniormost U.S. person in Libya.

The room for this meeting is deeply fortified, with exceptional control mechanisms limiting access to very few people. It is essentially a box, inside a compound, that is highly secure and impenetrable to electronic surveillance. It was a rustic and small space meant for high-level discussions of intelligence. Maps covered the walls and electronics were strictly forbidden inside the room. I had been in several of these through my travels, and you get a sense you're locked off from the rest of the world. The walls were white and yet the room was well worn. My staff person brought in a bottle of water, and the briefing was going to begin. I met the seniormost U.S. person and behind him Jeremy Freeman tried to join us.

Outmaneuvering the Deep State

In retrospect, our luckiest break in Libya was the opportunity to capitalize on a critical State Department mistake.

My staff person made it clear that I was to get the highest-level briefing possible, one that required the most secret clearance. I chimed in and insisted nothing be held back. I traveled from the United States, I was the chairman of the Subcommittee on National Security, four of America's best had been murdered in a terrorist attack roughly three weeks ago, and I wanted to know every detail I possibly could.

Freeman was asked verbally what level of security clearance he held. This was compared to the paperwork submitted in advance. While he had a security clearance, he didn't have a high enough clearance for a discussion at the security level we had come to receive. It wasn't my decision. It wasn't the decision of the person giving the briefing. It was the fact he didn't have the right clearance for this meeting.

Well, our blond-haired spy became irate. He began raising his voice, insisting and insisting again that he be allowed in the meeting. He told us that the top people in the State Department required he be in all discussions! This led to ten minutes of back-and-forth debate and argument about the reality of a security clearance. In a fit, he finally left the room and insisted that we not begin the discussion until he returned. That was a ridiculous request. My time there was limited, and I wasn't going to wait for the State Department to clear up some paperwork so they could listen in to what I was hearing. I was conducting an investigation.

Upon his return to the room, Freeman was more worked up than ever, insisting that we not have the briefing and demanding that his presence be granted. There was no possible way I was going to limit the information I could receive so that this guy could join the conversation when his only goal was to provide a report back to the seventh floor of the State Department.

It was extremely disruptive, and it slowed us down. We finally closed the door. I got the briefing and learned a lot. I wish I could share every bit of it with you but obviously it was highly classified. Most of it came out publicly; some critical components did not.

So here was the Deep State in action. About ten to fifteen minutes later there was a pounding at the door. We stopped the briefing to find Freeman demanding the briefer take a call from Cheryl Mills. A lawyer, she was the chief of staff to Secretary Clinton and one of the secretary's most trusted advisors. Cheryl Mills is legendary as Hillary Clinton's enforcer. Her tenure goes all the way back to 1999, when she defended President Bill Clinton during impeachment.

She was being called in the wee hours of the morning in Washington, D.C., to inform her of what was happening! Nobody I've ever spoken with has ever had such an experience. They evidently had a lengthy discussion because it took a long time for the briefer to return.

There was absolutely no reason for Jeremy Freeman to join us in this briefing, quite aside from him not having the proper security credentials. Congress has the duty and obligation to investigate matters in an unfettered manner. I was in Libya, and the State Department was trying to interfere. Their actions were meant to intimidate, suppress, and affect what I was hearing and seeing. His presence was making it exceptionally difficult for those on the ground to speak with candor, confidence, and clarity. As if it weren't enough that Congress was poking around, here was a young but senior bureaucrat from the agency they work for taking notes about what they were saying.

For example, I was trying as we toured the grounds of the compound to get the regional security officer (RSO) to share insight into what was happening. He whispered back that he was in no position to talk with other people around and it made him nervous that I would even stand close to him. I slid him my business card with my personal mobile phone and email address on it, hoping he would contact me at some point. Later we tried to bring him in for a transcribed interview in Washington, D.C., but he had some sort of breakdown and was unable to testify. We did not make a public note of this at the time, but it does show the severity of the pressure.

I thought there was contention before that meeting, but now the minder was enraged. We continued the tour and discussion until we needed to move to the airport.

As we swiftly darted out of the embassy and toward the airport, there were a few cars outside with people taking video of our convoy. This highly concerned our security detail and we drove as fast as we possibly could to the airport in the same dramatic fashion that brought us there. We boarded the plane and safely took off toward Germany.

As a side note, as we approached the German airspace it was now dark. Near approach, the plane had slowed and was circling when the pilot came back to tell us they were having problems with the landing gear.

Here we had a four-star general and a member of Congress on a plane trying to approach, and the landing gear wasn't going down into a locked position. Every emergency vehicle you could possibly imagine was lining the runway when the pilot said he had to take a chance and try to land the plane. They could not get confirmation that the landing gear had fully locked and there was a fear that when the wheel hit the runway, the gear would fold back up into the plane. General Ham said he had never been involved in such a thing and neither had I. I've traveled a few million miles in airplanes and I'd never been through that. Whew. All was well.

Getting the Backstory

Upon returning from Libya, I knew we weren't finished digging. But our efforts to get the whole story were met by a sluggish bureaucratic response—a classic Deep State tactic.

After our return, I found myself behind a dumpster at a Goodwill store in Virginia, hoping to meet an informant. Having recently returned from Libya, I had fresh perspective and information that needed to be exposed. Very few others knew what I knew.

My small but tenacious staff on the Oversight and Government Reform Committee's subcommittee on national security was willing to aggressively pursue the facts. We were trying to cultivate a potential whistleblower who had firsthand information about the security situation in Libya prior to the terrorist events on September 11, 2012.

Many of the people who had served in Libya during the attacks were being reassigned to other international posts, conveniently putting them out of the reach of congressional investigators. We knew of one who was in the United States for several weeks in language training school.

Eric Nordstrom had served as the regional security officer, commonly referred to as the RSO, in Libya prior to the attack. He departed Libya in August 2012, a few weeks before the bloody event in Benghazi, but had the firsthand knowledge of the security failures that played a role.

Tom Alexander, the staff director for the subcommittee who was trying to communicate with Nordstrom, noted that Nordstrom was nervous, very nervous. He knew very well that by telling the truth he could be risking his livelihood—or worse. He was trying to keep his head down and do his job. Speaking with Congress came with great personal peril and would likely suppress his future career progress. However, his testimony was pivotal to understanding why Benghazi happened.

We already knew from our trip that during his time in Libya, Nordstrom's superiors in Washington, D.C., had made it clear his requests were unwelcome and problematic. He came to understand that his persistence was unwanted. The Department of State didn't want to hear about security vulnerabilities. It was my understanding Nordstrom would potentially share details with us, but he was unsure whether to trust us.

Knowing the location of the language training school in Virginia, we found a retail store. It was a Goodwill store near the school. We passed along word that on a specific date Tom and I would be behind a dumpster nearby. We described the vehicle and indicated we would be there for an hour. Nordstrom could look up my picture on the Internet to verify my identity.

It was a beautiful fall day with a bit of crispness in the air. Tom and I parked close and wandered over to the dumpster and waited. I wasn't quite sure what I was going to say if someone from the store or law enforcement showed up to inquire, but I also knew we weren't doing anything wrong. In fact, we were doing the right thing pushing to get the truth and hopefully building confidence in a potential witness.

Part of me wondered if we were being set up. We were two guys in ties pulling up in Tom's two-door Mercedes-Benz and hanging out by a dumpster. It certainly did not look natural. It looked like a drug deal.

Nordstrom didn't show up that day but the fact that we made the effort was an important part of winning his confidence. He later told us why he didn't show. He had consulted with Legislative Affairs at the State Department about talking to us. Although their job is ostensibly to facilitate such communication, they told him in no uncertain terms that he absolutely should not meet with us. They told him he *could not* meet with us unless they were present. This is categorically false. As a whistleblower, he has every right to meet with us, and they have no authority to stop him.

In fact, they violate the law by trying to prevent him from coming to us.

But who is going to enforce that law? The Obama Justice Department? Attorney General Eric Holder? There are no consequences when the Deep State violates the law, because enforcement has become selective. These decisions are now political. The Deep State does not protect the whistleblower. And employees know very well that if they step out of line, they will be given some ridiculously bad assignment. Nordstrom was understandably protecting himself, his family, and his career. I respect that.

We were serious about getting to the truth and doing the right thing. Through some good staff work and work with him on the phone we eventually earned Nordstrom's confidence. We had some help from his former colleague Lieutenant Colonel Wood, the constituent who had first alerted me to the inconsistencies of the attack narrative. Irate with the knowledge that Legislative Affairs had prevented Nordstrom from talking to us, we demanded he be made available for a transcribed interview with our staff. We were able to document that he was in the country, leaving them no excuse to withhold his testimony.

When Nordstrom came in for his interview, the State Department came with him. Their presence tends to suppress candid conversations, which is the point. But to his credit, his public testimony a few weeks later was both honest and heartfelt.

In October 2012, Eric Nordstrom would join Lieutenant Colonel Wood, Ambassador Kennedy, and Deputy Secretary of State Lamb in a hearing before the Oversight and Government Reform Committee.

We learned that Lamb had essentially let them know: Don't ask for things. You're not going to get them. They wanted no paper trail, and they wanted no problems. Just do as you're told. So rather than react to the local security concerns by those on the ground, it was the classic Deep State Washington, D.C., signal that we know what is best. It was more of a political calculation than a security calculation.

Nordstrom, near the end of the hearing, would memorably describe the State Department this way: "For me, the Taliban is inside the building."

In the months after this important, initial hearing I would get to know the heroes who defended the personnel in Benghazi. Four died, but others lived. Kris "Tonto" Paronto, John "Tig" Tiegen, Mark "Oz" Geist, and Dave Ubben were heroes that night and are still my heroes today.

What the book and movie *13 Hours: The Secret Soldiers of Benghazi* couldn't share is what happened when the plane finally departed Benghazi carrying the shell-shocked Americans who had defended our consulate through the night.

They had literally been in the fight of their lives and saved more than thirty others from certain death. When the plane departed, it flew to Tripoli where they received initial medical care.

It never became clear why the plane went to Tripoli rather than Sigonella, Italy, where we had NATO assets, American personnel, and modern medicine. It would have been almost the same time and distance to go to Italy rather than Tripoli, where our American personnel had fled the embassy because of a pending attack.

Nevertheless, an American military plane was eventually dispatched to Tripoli to pick up the wounded Americans and take them to Germany. Dave Ubben in particular was severely injured.

Once patched up, the American heroes wanted to go home. Ubben, of course, could not. The State Department told them they were free to go but offered them no transportation to return home. No military airlift, no airline ticket, nothing. They were required to get their own tickets at their own expense to get home to their families.

To add insult to injury, the Department of State revoked their security clearances. This is how the State Department treated the heroes of Benghazi. This is how the Deep State suppresses, intimidates, and mistreats people. The Deep State didn't want them home talking about their experiences, and they certainly didn't want them to work again in their field of expertise.

Oh, and one more thing. Jeremy Freeman, the State Department employee who called Cheryl Mills and did all he could to thwart our investigation? Today, he still works for the State Department. That's the Deep State.

The Benghazi investigation almost didn't happen. Had I not heard from Lieutenant Colonel Wood, I wouldn't have traveled to Libya in time to get the unfiltered truth. We would have all been left to believe it was as Susan Rice said—a consequence of a YouTube video gone awry. There can be no doubt now that that was a lie.

More important, without exposing Benghazi we might never have learned that Hillary Clinton was using her private email server to conduct government business and transmit classified information. Benghazi was a symptom of a much deeper problem at the State Department. Their decisions were based not on a security calculation, but on a political one.

Even More State Department Shenanigans

The second time I encountered a "minder" from the State Department was during a February 2014 visit to the U.S. embassy in Papua New Guinea. I was there to learn more about how changes in the embassy design process would impact cost and security for the construction of a new embassy in Port Moresby.

As secretary of state, Colin Powell had developed a process called Standard Embassy Design, or SED, which used a secure design template for each embassy, saving both time and money. It meant construction didn't require reinventing the wheel each time with each new embassy. But when Hillary Clinton took over the State Department she worked closely with Senator John Kerry, and they reconfigured how the United States builds embassies. Clinton's plan, called Design Excellence, emphasized openness and "customized artistic designs." Right.

Within the bowels of the Department of State is the OBO, or Overseas Building Operations. Having reviewed a combination of inspector general (IG) reports and Government Accountability Office (GAO) audits, we found that the OBO seemed to be building embassies more slowly and more expensively than Powell's State Department did under his SED plan. This was costing taxpayers hundreds of millions of dollars and slowing down the ability to secure embassies around the world. We were building fewer embassies and spending more, which leaves more people at risk. Some of the older embassies are in bad shape and are not secure.

After seeing what happened in Libya, I knew this lack of security was serious and needed to be addressed quickly. So this investigation took me to Papua New Guinea. Yes, Papua New Guinea.

I was also deeply concerned about the incredible waste of money and time in construction of our embassies. The State Department had started building an embassy in Port Moresby only to stop its construction partway through. They decided to expand and redesign the building, which meant starting over.

Now, you should know a little bit about Port Moresby: It is the largest city in the South Pacific outside of Australia and New Zealand. It has a 60 percent unemployment rate. It is filled with poverty, shantytowns, and serious crimes like murder and rape. The Economist Intelligence Unit's Democracy Index rates it No. 5 on the list of ten least livable cities in the world.

The place is a dangerous horror, and Westerners are targeted.

So, the new embassy design would cost a few hundred million dollars more than the first design. Or, as CBS News reported, "The project estimate has ballooned from $50 million to $211 million, and according to an internal State Department document, there has been a 'termination of the current work and shuttering of the site until a new construction contract is awarded.'"

This was an incredibly dangerous city. There was absolutely no doubt in my mind that we needed a new embassy sooner rather than later. The existing embassy was in an extremely vulnerable situation because it was co-located with a bank that overhung the street parking, making both buildings an easy target for terrorists.

The State Department knew of our investigation. When I arrived, the ambassador had not been given prior notice that the new embassy was going to be dismantled in favor of a new design. I had to be the one to tell him because the State Department hadn't bothered.

State's decision was going to set back the opening date by several years. It was clear the department was communicating very little with anyone in that very small embassy.

Just like the situation in Benghazi, when I showed up there was an unexpected State Department official in the room. This time I was much more comfortable pushing back aggressively against people from Foggy Bottom. I asked the ambassador privately if he knew this gentleman. He said no, that this guy had shown up unannounced the day before and said he had flown in from D.C. and decided to sit in on our meeting.

We were meeting in a conference room where roughly eight of us sat around a table. I decided to go around the room and have each person introduce himself or herself. When I got to the Deep State mystery man, he had the gall to tell me that his presence was merely coincidental to my visit. He said the embassy building was his responsibility in Washington, D.C., and he just happened to think that it might be a good idea to go out and visit.

Keep in mind, he had never visited that city previously, nor was he communicating on a regular basis with the ambassador. I challenged the notion that his visit was coincidental. I said, "I'm going to give you one chance to tell the truth. And that will determine whether or not you stay. Are you here because I'm here?"

He responded, "Oh, no, I would have been here regardless of your visit. It was just a good time for me to come introduce myself to the staff at the embassy."

At that point, I unleashed on him and called him out for lying.

The State Department was given notice a few weeks ahead of my trip. They knew exactly when I would be there. I said, "You can get back on a plane and fly back to Washington, D.C. This is a total waste of taxpayer money. Your presence is not needed here and I will not allow you to sit in on this meeting." He looked over at the ambassador and it was clear he supported my decision.

The embassy regional security officer helped him get on a flight back to D.C. By the way, this flight is about eighteen hours, with two stops, about as far as you can get from Washington, D.C. Later the ambassador and support staff thanked me for highlighting how obnoxious he was in presenting himself as merely a coincidence. They were shocked and dismayed that State had not been candid about their situation.

This was just another instance where the State Department was doing everything possible to spy on a member of Congress and put a person in the room who could report back and help them manage the narrative.

I'm not aware of any other situation where an investigative body has been so manipulated and in such a brash fashion.

CHAPTER 10

Flouting Subpoenas

By now you've gotten a sense of some of the battles I had in Congress against the Deep State. You've seen some of their faces, know some of their names and their misdeeds. The worst of the federal bureaucrats try anything to evade sunlight.

As I began my time as a member of Congress, and later chairman of the House Committee on Oversight and Government Reform (OGR), I always assumed the most potent weapon in exposing corruption, incompetence, and other wrongdoing would be issuing a subpoena. Remember, as oversight chairman I could issue a subpoena for anything, on anyone, at anytime.

There are two types of subpoenas we would normally require. One compels the agency under investigation to produce documents. The other compels witnesses to testify. Only two exceptions to this obligation are recognized by Congress. First, when the president himself invokes executive privilege, indicating the information is part of the chief executive's personal conduct of his government duties, a subpoena can be legitimately blocked. The second allowable exception is when the information requested is part of certain grand jury material. Other than those two exceptions, Congress has a right to see everything. In theory. But in practice, not so much.

As you might imagine, the Deep State has developed innumerable strategies to play their version of Subvert the Subpoena.

Document subpoenas were more likely than testimony subpoenas to be subverted. I honestly don't know if we ever got 100 percent document production on any case. My OGR colleague Trey Gowdy often said, if you want 100 percent of the truth you need 100 percent of the documents.

The Deep State resistance against public disclosure of documents is a battle they must fight on two fronts. Not only do they have to deal with document subpoenas from Congress, but they are also required by law to produce documents requested by the public through the Freedom of Information Act. They use a similar arsenal of tactics to evade both types of disclosure requests.

One of the most common practices involved manipulating the media into reporting compliance with a document request simply because agencies had turned over "thousands of pages" of documents. Agencies would always tout the total number of documents they turned over to us. I wondered if the bureaucrats held stock in Staples or Office Depot. To inflate the page count, they would print copies of their agency websites. They would provide news and magazine articles. If they had a back-and-forth exchange among multiple employees, they would print the same exchange from each separate email account. No matter how long that exchange was, we'd get multiple copies of it, and they would count them as separate documents and pages.

Then they'd claim to reporters they'd already turned over thousands of pages, making it sound like they were wearing out their printer to provide documents, even as they withheld the documents that would have exposed the truth.

For example, when we were investigating the ATF "Fast and Furious" scheme, more than three years after we started our investigation, on November 4, 2014, *Politico* reported, "The Department of Justice has turned over nearly 65,000 pages of documents on the 'Fast and Furious' gunrunning scandal to the House Committee on Oversight and Government Reform." Sounds like a lot, right? Sounds like we might be getting some really meaty information about what went on at the ATF and the DOJ during that operation, right?

Well, as *Politico* elaborated, "The tens of thousands of pages of documents released to the House late Monday include an email exchange between [Eric] Holder and his wife. . . . A person familiar with the exchanges between Holder and his wife, Sharon Malone, said that one involves Malone forwarding a news article related to the case. In the email thread, the couple went on to discuss family matters not related to the Fast and Furious investigation."

Let's just say that particular document dump wasn't helpful.

Another common tactic is to produce a document and then to redact or black out close to 100 percent of the page. A classic instance played out during "Fast and Furious." CBS News ran a story on April 27, 2012, showing their viewers documents they received more than a year after requesting them through the Freedom of Information Act (FOIA). The CBS FOIA request got the same kind of response Congress was getting on document subpoenas. The response, said CBS, included "mostly-blank pages."

CBS uncovered a fascinating detail. One of the DOJ officials they requested documents from, Kevin Carwile, chief of the Capital Case Unit, sent an email on February 1, 2010, vowing, "I haven't forgotten you . . . I will call you in the morning." To whom he sent that email and what they might discuss when he finally did call, no one outside the DOJ knew.

The FOIA laws used by CBS to uncover documents are important. According to the website FOIA.gov, "Since 1967, the Freedom of Information Act (FOIA) has provided the public the right to request access to records from any federal agency. It is often described as the law that keeps citizens in the know about their government." Watchdog groups like Judicial Watch have had some success in getting information by using FOIA, but Deep State dragons have figured out that they can hamper those requests by abusing exemptions to the act.

We ran up against that strategy when we were investigating the TSA in May 2014. The OGR was given a report with the unwieldy title "Pseudo-Classification of Executive Branch Documents: Problems with the Transportation Security Administration's Use of the Sensitive Security Information (SSI) Designation."

In it, we were flat-out told by Andrew Colsky, the Sensitive Security Information director for the TSA, that "currently I sit in the Freedom of Information Act office. And one of the first things I was told when I got there from both attorneys and FOIA processors was, oh yeah, don't worry about it, because **if you come across embarrassing information or whatever, [the chief counsel] will just hide it and come up with an exemption; because if you cover it with a FOIA exemption**, it's so hard for the other person to challenge it, and it will be costly and difficult for them to challenge it, and they're probably never going to see it anyway, so you just get away with it. That's the way it's done." (That bold-faced admission is lifted directly from the report, by the way.)

Colsky, as the SSI office director, would know. His department, according to the TSA website, oversees "information that, if publicly released, would be detrimental to transportation security."

In other words, SSI decides secrecy based on whatever the TSA says it is. And here we have a guy who knows what he's talking about flat-out telling you how the Deep State gets around our congressional duty to provide oversight.

Standing Up to Congress

Agencies play a whole other series of games when they receive a congressional subpoena for witness testimony.

First, witnesses refuse to show up when they're scheduled to testify. The most egregious case of that tactic I ever saw was that of Bryan Pagliano. Having served as the IT director of Hillary Clinton's 2008 presidential campaign, Pagliano was the specialist Hillary tagged to set up the private email server. He set it up in the basement of her home in Chappaqua, New York, in 2009 once she became secretary of state. When the House Select Committee on Benghazi held hearings in 2015, Pagliano was someone they wanted to talk to so they could track down all Clinton's emails relating to the attack.

Chairman Gowdy issued a subpoena for Pagliano's testimony. His lawyers sent a letter back stating that the IT guy would assert his Fifth Amendment right not to testify. Pagliano was scheduled to appear before the committee and be deposed on September 10, 2015, but instead met behind closed doors with committee members for about twenty minutes. Afterward, ranking member Elijah Cummings told reporters, "I don't think he has any information about Benghazi," according to a CBS News story at the time.

The following March, Pagliano and his legal team cut a deal with the Department of Justice so he could cooperate with the FBI in return for immunity from prosecution in any criminal investigation into Clinton's potential mishandling of classified information. His lawyers tried and succeeded in keeping the details of the immunity agreement private.

We subpoenaed Pagliano to testify before the OGR hearing on "Examining Preservation of State Department Records" that I chaired on September 13, 2016. Now, even though he had already talked to the FBI, and despite his immunity deal with the DOJ,

Pagliano wouldn't talk to us. In fact, he just plain did not show up! The appointed hour, 10 a.m., came and went. No Pagliano. His lawyers said he was pleading the Fifth.

In my opening statement before the hearing, I said, "I take my responsibility as a committee chairman seriously, especially the decision to issue a subpoena. It's a serious matter. Mr. Pagliano has chosen to evade a subpoena, duly issued by the committee of the United States House of Representatives. I will consult with counsel and my colleagues to consider a full range of options available to address Mr. Pagliano's failure to appear." Minutes later, to Elijah Cummings I added, "This is not an optional activity. You don't just get to say, hey, well, you know, I decided not to do that. . . . If anybody is under any illusion that I'm going to let go of this and just let it sail off into the sunset, they are very ill-advised."

We issued a second subpoena for Pagliano to appear on September 22. This time we had the U.S. marshals serve it. Once more he was a no-show. So we passed a resolution that found Pagliano in contempt of Congress. My Democrat colleagues would have the public believe that our purpose was to embarrass the Clinton campaign a few weeks before the presidential election on November 8. Not so. While they were focused on the politics of the moment, I wanted to underscore the fact that Pagliano's refusal to honor the subpoena undermined the authority of Congress to provide oversight. "This committee cannot operate—it cannot perform its duty, nor can any committee of Congress—if its subpoenas are ignored," I said at the time.

In fact, I felt so strongly about the matter that on February 16, 2017, I sent a letter to President Donald Trump's attorney general, Jeff Sessions, asking for Bryan Pagliano to be prosecuted for failing to appear before Congress. In the letter, I reiterated my point that "[i]f left unaddressed, Pagliano's conduct in ignoring a lawful congressional subpoena could gravely impair Congress's ability to exercise its core constitutional authorities of oversight and legislation." Despite my request that Pagliano be brought before a grand jury or for him to be charged with violation of U.S. Civil Code 192, a misdemeanor that would have seen him fined a maximum of one thousand dollars and jailed for up to a year, no charges were brought.

As I said many times on Fox News throughout spring and summer of 2018, I believed Attorney General Jeff Sessions should resign. He is a decent man, but he is just not strong and determined enough to combat the forces trying to destroy President Trump.

The problem with this incident is that it set a precedent. In fact, in March 2018 I saw several YouTube and other online references claiming that former Trump aide Sam Nunberg would refuse to cooperate with Special Counsel Robert Mueller in his investigation into Russian interference in the 2016 presidential election, just "like Bryan Pagliano during the Clinton FBI email probe."

The fact that Pagliano suffered no consequences for his intransigence underscores how little regard bureaucrats have for a congressional subpoena.

The Deep State has developed several other ways to subvert justice when it comes to subpoenas. One is to send someone to testify who doesn't know the answers. Another is to send someone to testify who is thoroughly schooled in the art of evading the question.

Now, within the executive branch of the federal government—the president's branch—is a department known as the Office of Management and Budget (OMB). Its mission is to "serve[s] the President of the United States in overseeing the implementation of his vision across the Executive Branch." That also means it oversees federal agencies.

OMB happens to be the clearinghouse for witness testimony, too, it turns out. They practice with people we subpoena, teaching them and coaching them on their testimony. Think about it: OMB uses millions of taxpayer dollars to coach somebody on how to withhold the truth. It's not like OMB is telling them to be honest, forthright, to tell it like it is. There is a legislative liaison whose job is to *make things go away*.

Told Not to Come

Take the case of Ronald Turk, for instance. Turk was the associate deputy director and chief operating officer at the Bureau of Alcohol, Tobacco, Firearms and Explosives (ATF). We were looking into various ATF failures that ultimately led to the death of Immigration and Customs Enforcement (ICE) agent Jaime Zapata back in a 2011 shoot-out with the Mexican drug cartel, Los Zetas. It seems that the ATF had been tracking two potential weapons traffickers at various Texas gun shows and pawnshops and had an opportunity to seize the firearms they purchased and arrest the wrongdoers. But they didn't do so. Those guns killed Zapata.

We held a hearing on March 9, 2017, and Ronald Turk had been subpoenaed to testify. Like Pagliano, he never showed up. A month later, on April 4, we held another hearing. This time the subject was the use of confidential informants at ATF and DEA. Turk was scheduled to appear here and did. I asked him why he neglected to come to the previous hearing.

Turk started off claiming, "We had about, I want to say eight days' notice with that hearing. Right out of the box, we were advised by the Department of Justice that they thought there were certain, for lack of a better word, etiquette rules from past committees . . . for things like fourteen-day rules." Well, there is no such thing as a fourteen-day rule and never was. We don't have to give you two weeks to prepare for your congressional appearance.

Then he tried to convince me that he thought accepting an invitation to testify before Congress was nonobligatory. He said he had discussed it with the Department of Justice, which gave him and another ATF official "guidance" that they did not need to come before our committee and answer questions.

I started by taking the opportunity to set him straight, saying, "Listen, we appreciate you being here today. I appreciate what you do for this country. But attending a congressional hearing is not an optional activity. . . . And for you to . . . unilaterally just kind of collectively say, well, two of our people aren't going to show up . . . come on, that doesn't happen in any other setting. And you should be ashamed of yourself for that."

Mr. Turk took offense. We went back and forth for a few minutes.

TURK: Well, sir, I would say . . . you're questioning my honor now.

CHAFFETZ: Yeah, I am.

TURK: I don't appreciate that.

CHAFFETZ: You were invited to come. . . .

TURK: I was following guidance. I'm a good soldier. I was following orders.

By this time, after years of investigating the ATF's gun-walking schemes, I recognized this tactic. Now that I had Turk before me, I decided to find out names.

"Who told you not to come?" I demanded to know.

Turk responded: "The Department of Justice."

That sure wasn't a good enough answer. "Who? I want to know a name," I countered.

"There's so many names I could probably list off eight names between the—"

I cut him off. "Good," I said. "Start. Give me one."

I was determined to find out who was behind the obfuscation. "Who all was this discussed with?"

He stalled, trying to decide if I meant I really wanted to know whom he talked to.

"Yes," I clarified. "Who told you not to come? Start naming them."

"Sir, are you suggesting I shouldn't follow . . . guidance from the Department of Justice?"

"I want to know who's giving you that guidance. You told me there were at least eight [people], and that you could name more, so give me one and then we'll start with number two."

Turk finally gave a name. "Sure," he said. "Mr. Ramer. He's the head of the Office of Legislative Affairs."

Finally! A head from the Deep State had popped up in the middle of a congressional hearing. This one happened to be Sam Ramer, the acting assistant attorney general for the DOJ's Office of Legislative Affairs. I kept at it until Turk gave me about four more names and then promised to get me the final three. But he was not happy about it.

"Sir, I came here as a good, honorable person and you, you admitted that you challenged my honor. I do not appreciate that. I operated in good faith—" he began. I had to interrupt him. "Operated in good faith?" I said. "You didn't show up!" The exchange got a little more heated.

TURK: Sir, I was told that it was an invitation as [ATF] director [Thomas] Brandon explained. I did not know that that was not optional. . . . I would have happily gone against Department of Justice guidelines and been here that day. . . . Everyone I've ever talked to about this understands precisely what happened that day. Yet you want to get your fifteen seconds of YouTube minute time to challenge my honor.

A YouTube minute? I don't think so.

CHAFFETZ: I'm not here to disparage your entire career based on one incident. I'm here to say that that one incident was a really, really bad decision. And we're tired of people saying, well, I'll brief staff or I'll talk to you privately, when we're trying to do it in the open light of day. We have very valuable time and lots of important things to deal with here. That's why when you are invited to Congress you're expected to show up. . . . And for you to suggest that as the chairman of the committee I'm just here to get a YouTube moment . . . are you kidding me?

TURK: Sir, I can assure you in the future—

CHAFFETZ: You don't think . . . that [what] we're trying to do on these . . . cases is a value to the American people? When you hide information . . . you don't provide it to the United States Congress . . . and we can't get the answers to the questions that we have. There's a reason why I do have to issue subpoenas. And these two agencies, DEA and ATF, I love the men and women who do this. But the management I got a serious problem with. . . . That's why we're doing these types of hearings—is because you do need to be held accountable. And when you're invited to Congress you don't sit around and have a group and say, well, it's probably in our best interest not to show up. We act in the best interest of the American people and you don't respect that. That's why you didn't show up.

TURK: Sir, I do respect that tremendously. I can assure you that in the future regardless of what guidance I'm given from the Department of Justice I will be very responsive to this committee. I can guarantee you that.

It's Never About Your Name in a Headline

We may have forced one member of the Deep State to give up his secrets, but there are so many more. To cut through all the red tape, blockades, and impediments various governmental agencies flung in our way, we have had to develop a few different strategies.

It wasn't always necessary to convene a congressional hearing to get the information we needed. Sometimes we just needed to interview a cooperating witness under oath. On other occasions we had friendly witnesses who asked to be subpoenaed to give them cover with their boss. I obviously can't share those names. But they could tell their boss they had no choice, they had to tell us. They were very grateful for the cover of a subpoena. It was an effective tool.

We could also talk directly to a witness. I remember, after James Comey announced he wasn't going to prosecute Hillary Clinton over her emails in July 2016, I was able to get him on the phone that same day. Our conversation went something like this:

CHAFFETZ: We need you to come up to the Hill and testify.

COMEY: Yes. I understand.

CHAFFETZ: Would you like me to issue a subpoena—would that help you?

COMEY: No, no, I'm happy to come up. In fact, don't issue a subpoena. I would really rather you didn't.

So, we didn't. I asked which day was most convenient for him, he gave me a date, and that's when we held it.

Comey knew he had to come up. He understood. I was very grateful and appreciative that he was so willing to give so much time to come up and answer our questions. On some of the most important questions he was very elusive, though. In retrospect, he was not so candid. He is a professional at testifying. Those are the people the Deep State likes to send to hearings.

When you have $4 trillion flowing out of the federal budget, there are a lot of incentives to be discreet . . . and secretive.

We have only seventy people on the staff of the OGR and there are two million federal employees. There is a lot we still haven't uncovered. We can't react to every newspaper story. We're not just there to chase headlines.

The OGR has a whistleblower hotline that generates twenty to fifty calls a day. Right on our home page is a big, bold headline that says **BLOW THE WHISTLE.** When you click on it you are requested, "Please use the form below to alert Chairman Gowdy to fraud and abuse in your agency or other organization."

Many of the alerts we get come from frustrated employees mad at their boss over typical workplace activities. Every once in a while something comes across our desk that is much more serious. Those whistleblowers give us a road map to root out corruption, sexual harassment, and perilous or just plain criminal schemes.

Frankly, I have to say that these courageous women and men, who risk not only their livelihood but often their physical safety as well, were perhaps the most effective weapon we had in combating the Deep State. If it hadn't been for ATF agent John Dodson, we might not have heard of Operation Fast and Furious. After repeated complaints to his agency's Office of Professional Responsibility were ignored, Dodson went to Senator Charles Grassley and told him that American gun dealers were selling arms to Mexican drug cartels at the ATF's request.

Dodson faced serious blowback from his agency after exposing the gun-walking scandal in 2011. While he is still an ATF agent as of this writing, he has been retaliated against, transferred, and marginalized, according to a 2017 story in the *Daily Signal*. Looking back six years later, he told investigative reporter Sharyl Attkisson on her syndicated program *Full Measure* that in going public, he "went from being an agent of the government . . . to an enemy of the state."

Hurting Whistleblowers and Hiding It

He's not the only one. The TSA, for one, has been especially hard on whistleblowers. In fact, they lost a 2015 Supreme Court case over their treatment of Robert MacLean, an air marshal who exposed several questionable TSA policy decisions. The TSA fired him in 2006 for talking to an MSNBC reporter about "a Department of Homeland Security plan that would have reduced the number of armed marshals on commercial aircraft at a time when intelligence officials were warning of an imminent al-Qaeda hijacking threat," said the *Washington Post* in a profile of MacLean on March 3, 2016.

MacLean was dismissed allegedly for leaking sensitive security information to the media—even though that information was only deemed classified three years after MacLean's initial disclosure.

This wasn't the first time the TSA had been taken to task over their use of sensitive security information (SSI). You remember what Andrew Colsky told us about the way the TSA hides information. Even the agency's inspector general, John Roth, was frustrated with its secrecy. In his December 30, 2016, report to the Department of Homeland Security, Roth complained about the TSA redacting large chunks of material from critical records when he was trying to investigate ways the TSA could improve security at airports.

"The redactions are unjustifiable and redact information that had been publicly disclosed in previous Office of Inspector General (OIG) reports." He goes on to say, "I can only conclude that TSA is abusing its stewardship of the SSI program. None of these redactions will make us safer and simply highlight the inconsistent and arbitrary nature of decisions that TSA makes regarding SSI information. This episode is more evidence that TSA cannot be trusted to administer the program in a reasonable manner."

Roth couldn't have made our point more clearly. When we conducted the "Transparency at TSA" hearing on March 2, 2017, we were trying to get to the bottom of whistleblower retaliation complaints at the agency. Carolyn Lerner was the special counsel in the U.S. Office of Special Counsel (OSC). According to the report she presented to us, that office is "an independent investigative and prosecutorial federal agency that promotes accountability, integrity, and fairness in the federal workplace. . . . And we protect federal employees from prohibited personnel practices, most notably whistleblower retaliation."

Lerner's report concerned "OSC's investigations of whistleblower retaliation complaints at TSA. [Since 2012] OSC has received more than 350 whistleblower retaliation cases from TSA employees."

She focused on four complaints in which "TSA officials . . . experienced involuntary geographical reassignments, a demotion, and a removal, all of which were allegedly in retaliation for protected whistleblower disclosures." Lerner sought documentation of the complaints from the TSA, with little success. The agency delayed sending her the papers, "asserting claims of attorney-client privilege," Lerner wrote.

Her report included an exhibit that clearly illustrated the problem. It was a piece of white paper with everything on it blacked out except the words "OSC Exhibit, March 2, 2017 (Provided by TSA to OSC on October 20, 2016)." Like CBS with the DOJ dossier, "TSA redacted every word of the document, including the date, author, and recipient," Lerner wrote.

Explaining the TSA's actions at the hearing was Huban Gowadia, the agency's acting administrator. I asked her why Lerner wasn't getting 100 percent of the information she needed to investigate the whistleblower retaliation complaints. Gowadia claimed that attorney-client privilege prevented her from disclosing everything, but she tried to assure me that she would give the Office of Special Counsel "all the appropriate information."

Pressing her, I said, "so what information do you believe the OSC should not see?" She tried to evade a direct answer. "I have to stress that TSA is not an agency independent. We follow guidance that the department gives us. I can assure you that we will follow up with this at the department level."

I was not about to let her shove this back under the TSA carpet. "You're relying on guidance from the department and you're going to withhold that information from Congress?" I said. Her reply shocked me. "To the best of my knowledge, the guidance is not in writing."

"So, wait a second," I interrupted. "You just made this up? It's not in writing?" I turned to Lerner, the attorney, to see if this was true. "There is no attorney-client privilege when one government agency is investigating another agency," she stated firmly.

Inventing excuses, hiding facts, creating delays to hand over subpoenaed information—this is classic Deep State behavior. We try our best to cut through it, but, as I remarked at the end of the TSA hearing, whistleblowers must "know the deck is stacked against them."

A Rare Bright Spot

One agency where whistleblowers were able to effect change was the U.S. Forest Service. For decades, women in that service had filed reports of sexual harassment, then filed reports detailing the retaliation that followed. Two class action lawsuits (the first as early as 1973) and two "consent decrees" that demanded the Forest Service change the way it treats women didn't make a difference. All these years later, on December 1, 2016, I found myself chairing a hearing titled "Examining Sexual Harassment and Gender Discrimination at the U.S. Department of Agriculture."

I listened as one of the whistleblowers, Denise Rice, a fire prevention technician in California, barely held back tears while she told us how her male boss had grabbed a letter opener, then "poked my breast, both breasts, with a smile on his face in an arrogant way, like he could get away with it." She detailed several other encounters that occurred over the two-year period from 2009 to 2011. "He has cornered me in the bathroom, he has lifted my shirt up. I would wait till everybody would leave . . . and he would be waiting for me. He called me constantly, he interfered with everything. He stalked me," she declared. When she complained, Forest Service management "removed all of my supervisory duties, moved me from my location, and isolated me," she said. And what happened to the boss when his violations were reported? He was allowed to retire, and then was actually brought back as a motivational speaker.

My colleagues and I were incensed. After hearing from several bureaucrats about how the agency has gotten better at dealing with sexual harassment and acknowledging whistleblowers, Trey Gowdy said, "Well, I just heard the most glowing account of all of the improvements that have been made over the past eight years and you mean to tell me that someone can engage in the conduct Miss Rice just described and avoid all consequence whatsoever?"

Lenise Lago, the deputy chief of business operations for the U.S. Forest Service, confirmed that "[p]er the federal regulations yes. Someone can retire or resign in lieu of being removed."

One of the "takeaways" from the hearing (you can look it up on the OGR website) was that "[h]arassment and discrimination at USDA has gotten worse under the Obama Administration. Witnesses testified sexual assault, harassment, discrimination, and resulting retaliation has increased at the Agency since 2008." At the end, I promised the Forest Service firefighters, "On behalf of all of the members we will go to the end of the earth to protect you and the other women that have gone through this. . . . To those in management and in other positions . . . we will use every power we possibly can from this pulpit to make sure that they are treated with dignity and never have to go through that again in any way, shape, or form. You will see more subpoenas and more hearings than you can possibly imagine if we hear one thing about any sort of reprisal in any way, shape, or form. I can't say that strongly enough."

We didn't need to hold any more hearings. After a comprehensive 2018 investigation by *PBS NewsHour,* where thirty-four women in thirteen states detailed incidents of rape, retaliation, "gender discrimination, bullying, sexual harassment and assault by crew members and supervisors," there was some positive action. On March 7, 2018, Tony Tooke, chief of the U.S. Forest Service, resigned amid allegations of sexual misconduct. Tooke had only been in charge for six months but he had worked in senior positions at the service for years.

Bottom line: Congress is going to have to stand up for itself and be more aggressive about enforcing subpoenas. And protecting brave bureaucrats who call out corruption *must* be a priority.

Challenging the very foundation of the Deep State is, honestly, a scary and even dangerous thing to do. Washington, D.C., does not want to be disrupted. There will be a price to pay. Just ask a man who has disrupted business as usual . . . the president of the United States.

What I call the Deep State—not just the traditional uses of the term meant to identify intelligence and military-industrial interests—but the true Deep State, which is the permanent class of Democrats, Republicans, federal bureaucrats, and entrenched Washington, D.C., and Acela Corridor insiders—is not going to go quietly. President Donald Trump is arguably the greatest disruptive threat to the Deep State we have ever seen. This is not a short-term fight.

As we will see in the next chapter, the Deep State is trying to weaponize everything in their power to destroy President Trump.

CHAPTER 11

The Deep State's Nightmare

The political earthquake that was the election of Donald J. Trump to be the forty-fifth president of the United States reverberated across the political spectrum, through the media, and across the globe.

Within days, my congressional office was being flooded with so many phone calls, my twelve-member staff could not begin to field them all. Few calls originated in my home district, and the vast majority read from the same scripts demanding investigations or tax returns. The panic from Clinton voters was palpable. Donald Trump was going to destroy America, they said. He would crash the economy, wreak havoc on the planet, target gay people, and start World War III. These were not the words of a few extremists. This was a common theme.

As laughable as these assumptions are in light of current events, the people espousing these views truly believed them. They wanted to convey the imperative that I, as chairman of the House Oversight Committee, open an investigation of the president-elect immediately to stop him from even being sworn into office. As implausible as this demand may seem, we heard it every single day.

Allegations against the incoming president flew fast and hard— and Clinton voters seemed to accept them all at face value. They wanted to know if I would go after Trump as hard as I had gone after Hillary Clinton and President Obama. The inconvenient truth that Secretary Clinton was never investigated by the committee until after she had stepped down as secretary of state was lost on them. The reaction to the election by the left was nothing short of hysteria.

The calls continued unabated for months. My constituents could not get through to get the help they needed. After a nearly suicidal veteran was unsuccessful getting through to our office for days, we finally took the drastic and unusual measure of accepting calls only from area codes in our district. All others went to voice mail for a time. During the first four months after the election, our voice mailbox, which could hold hundreds of messages, would fill up daily.

Although Clinton voters were beside themselves, those of us who had been toiling to get documents and evidence from the Department of Justice and other executive branch agencies were hopeful. Perhaps this president could help us break the Deep State impasse.

Expectations of the Trump Presidency

The night of the election, I expected (like most people) that Hillary Clinton was probably going to be elected the next president of the United States. I knew the Oversight Committee was going to have a major task ahead of us, given the number of scandals she had created. Between her unsecured server hosting classified secrets, her Uranium One deal with the Russians, and the many irregularities with her Clinton Foundation charity, I assumed her past actions alone would keep the committee busy for years—not to mention anything she was likely to do once in power.

Within days of the astonishing and unexpected victory of Donald Trump, I began to realize the implications of this election for our committee. We would finally be able to complete outstanding investigations, including getting to the bottom of the Clinton server scandal, finally getting the documents we'd been fighting for in the Fast and Furious investigation, and perhaps holding people at IRS accountable for their reprehensible targeting of free speech. I was euphoric about the possibilities.

The announcement that Senator Jeff Sessions would head up the DOJ was a welcome one. Our staff had worked well with Sessions's staff. As a member of the Senate Judiciary Committee, Sessions was on the record criticizing the FBI's investigation of Clinton's email case. A November 18, 2016, headline in the *Washington Post* read, "Jeff Sessions as Attorney General Could Mean Trouble for Hillary Clinton and Her Family's Foundation." Sessions told Fox Business Network's Lou Dobbs in an October 30, 2016, interview, "I'm uncomfortable [with] the way the investigation was conducted. I think it should have used a grand jury and people who were given immunity should have been taken before the grand jury then . . . and you grill them. Because they will be protective of the people they like and they work with. People will surprise you how they'll just spill the beans when they're under oath like that." In reference to FBI director James Comey's failure to charge Hillary Clinton, Sessions told Dobbs, "It did appear to me there was sufficient evidence to bring a charge" against Clinton.

Meet the New Boss, Same as the Old Boss

With a new sense of optimism, I asked for a meeting with Sessions a few weeks after he was sworn in as the new attorney general. I put together a professional presentation outlining all the previous document requests to which the DOJ had refused to respond. Finally, we were about to get the answers the public deserved. Here we had a Republican president in place. We had a new attorney general who had vocally supported these inquiries. I was granted an hour with the attorney general one morning in March 2017. I drove over to the Department of Justice with two Oversight Committee staffers and we met in the attorney general's rather large conference room. I had with me a large stack of requests that had gone unfulfilled by the DOJ. We started going down the list.

At the top of my list was the most egregious—and I believed the most easily resolved—dispute. The failure of Clinton information technology aide Bryan Pagliano to appear before the committee in response to a duly issued subpoena was unacceptable. We had issued a subpoena to compel his attendance, twice, and he had refused to appear before the committee. He could certainly plead "the Fifth" but to snub his nose at Congress and refuse to show up was outrageous and unprecedented. It could not stand.

This should not even have been a partisan issue. Even Democrats on the House Oversight Committee would agree that congressional subpoenas mean something. I wanted the Justice Department to prosecute Pagliano. Sessions refused: "No. I can't do that. I can't talk about it." The Pagliano issue wasn't even about Hillary Clinton—it was about whether congressional subpoenas hold any force at the DOJ. This was a black-and-white case. But here was a Republican attorney general telling me, No—we're not going to do anything with that.

As I went down my list, anything that had to do with Hillary Clinton was a no.

He would make no promises or assurances on any item on the list, even the ones that had nothing to do with Clinton or the State

Department. By the end of the hour, it became clear this Justice Department was going to be little to no help, same as the Obama administration.

The same Deep State that I was working with during the Lynch-Holder years was still there. Nothing had changed with a new administration. I couldn't see that Sessions was doing anything to drain the swamp within the DOJ.

I pushed hard with Sessions on the Pagliano issue. At the conclusion of the meeting, Sessions promised he would follow up with me, particularly on the Pagliano case because I would not let that go. I couldn't take no for an answer. I was infuriated that the fate of congressional oversight was in the hands of the attorney general.

When Sessions got back to me, I was in Springville, Utah, departing a meeting with a local Chamber of Commerce. I took the AG's call as I drove and pulled into the parking lot of a local supermarket to be fully focused on the discussion. I expressed to him again my deep frustration with this process. I told him Senator Sessions would never put up with this. As a judge, he would never put up with this. What had changed?

The Pagliano case was too close to Hillary Clinton. He refused to do anything with it. I couldn't believe it. I couldn't believe this was the position of the Trump administration.

I would later visit with then White House chief of staff Reince Priebus, White House general counsel Don McGahn, and even then–White House chief strategist Steve Bannon. I spoke to them about these requests, hoping they would feel differently than the attorney general. To my genuine surprise, it was clear that none of them was going to push DOJ. All three of them were receptive and sympathetic to what I was asking for, but they were taking a very hands-off approach. The result was that the Deep State got to continue to run things.

I went from the euphoric high of Donald Trump being elected in November to the crushing reality in March that nothing was going to change. Further, it was clear that neither Speaker Paul Ryan nor House majority leader Kevin McCarthy had any desire to fight this fight. Their focus was on other things. They didn't place oversight high on their priority list. Not even the other chairmen of House committees were on board. In particular, House Judiciary Committee chairman Bob Goodlatte, who could have been very influential on this issue, didn't want to do anything to rock the boat.

Combine that lack of enthusiasm with a media that had an insatiable desire to take down Trump and it was not a good scenario in which to make progress. I became disillusioned and deeply frustrated that Congress could not do its job. If we couldn't find the wherewithal to defend our power under these most favorable circumstances, when would we? These challenges contributed to my decision weeks later to leave Congress.

As of June 2018, Fast and Furious documents still aren't in the hands of the House Oversight Committee, despite seven years and a long court battle to obtain them. The Justice Department declined to prosecute EPA's head of the Chemical Safety Board, Rafael Moure-Eraso, whose case was a slam dunk. They never did prosecute Bryan Pagliano for a clear failure to comply with a congressional subpoena—a decision I believe will come back to haunt both parties in this Congress.

Remember, I was elected to Congress in 2008, the same year Barack Obama became president. I can tell you that the contrast between Obama's Washington and Trump's Washington is dramatic.

Although Donald Trump is a Republican, the fact is that he ran against both the Republican and Democratic establishments. His mission was to disrupt business as usual, and his nature is to keep everyone—allies and adversaries—guessing. Nobody can argue that he hasn't done that.

Perhaps this is one reason the president is treated completely differently than anything I saw in the previous eight years. For the Deep State, all the rules have changed. The constitutional priority to enforce the law seems to have been replaced with a prerogative to sabotage and undermine the duly elected president of the United States.

A Flood of Partisan Leaks

Washington has always had a leaking problem, but with the election of Trump the leaks became a deluge.

The leaking of classified or sensitive information intended to promote a political narrative is a problem that has dogged the Trump administration even before day one. Whereas the Deep State had seemed to function in lockstep with the prior administration, it became immediately clear there would be no such support for President Trump.

One of the most obvious examples of deliberate sabotage took place before he was even sworn in. FBI director James Comey and Director of National Intelligence James Clapper appear to have colluded with CNN to release information that would provide a pretext for the dissemination of the Trump "dossier." The dossier, filled with salacious and unverified gossip about Donald Trump and paid for by the Hillary Clinton campaign and the Democratic National Committee, fueled the political narrative that Trump had colluded with the Russians to defeat Hillary Clinton in the 2016 election. Despite breathless coverage of that conspiracy theory by CNN and other Democrat-friendly outlets, hard evidence has been sparse.

The release of the dossier in January 2017 ignited endless speculation, but the Deep State set up by Obama holdovers Comey and Clapper would not be known until Comey's book was released in April 2018, long after the damage was done. Comey admitted in his book that he told the president-elect that CNN had the dossier and was looking for a news hook. With the leaking of Comey's meeting with the president-elect, they got their news hook.

Subsequent leaks, designed to damage the president, targeted his appointees and casually compromised the safety and security of the United States and its allies.

Within a day of the firing of White House national security advisor Michael Flynn, the *Washington Free Beacon* reported that Obama loyalists had been plotting to take down Flynn to prevent him from revealing secret agreements that had been part of the Iran nuclear deal. One anonymous source told the *Free Beacon*, "It's undeniable that the campaign to discredit Flynn was well underway before Inauguration Day, with a very troublesome and politicized series of leaks designed to undermine him." One White House advisor who was also a member of President Trump's National Security Council at the time told the *Free Beacon*, "The larger issue that should trouble the American people is the far-reaching power of unknown, unelected apparatchiks in the Intelligence Community deciding for themselves both who serves in government and what is an acceptable policy they will allow the elected representatives of the people to pursue."

Some leaks have posed serious threats to national security. For example, last year an unidentified leaker revealed sensitive Israeli intelligence to the *New York Times*, which the *Times* reported in June 2017. In the haste to exploit any potential mistake by the new administration and hoping to embarrass the president for allegedly sharing the intelligence with the Russians, the leaker released and the *Times* published the classified secrets for all to read. An opinion piece on the English-language site (YnetNews.com) of Israel's biggest newspaper called the leak "an intelligence catastrophe." The author, Alex Fishman, wrote, "If there is even a grain of truth in the recent reports, it means someone is waging their war on Trump at Israel's expense, intentionally causing serious damage to America's ally—verging on treason."

Admittedly, we do not know whether current government officials or past Obama appointees are behind such leaks. The finger-pointing goes in every direction. However, we know the twenty-seven anti-leaking investigations currently under way by this administration are important and justified.

These leaks should never be mistaken for legitimate whistleblower activity that is intended to correct mismanagement.

In my work on the House Oversight Committee, I have come to value government whistleblowers. Many of them are patriots who want mismanagement to be dealt with in a system that tends to reward it. With two million federal employees, there's always somebody doing something stupid somewhere. Any federal worker with a legitimate complaint can bring their concerns to the Office of the Inspector General or Congress and receive protection from retaliation. We have a network of seventy inspectors general with a total staff of 13,500 whose job is to investigate wrongdoing within federal agencies. Whistleblowers can follow the proper procedure for addressing mismanagement.

Premature Calls for Investigations

When allegations of collusion between the Trump campaign and the Russian government first began to appear, many angry Clinton voters turned to the House Oversight Committee to demand investigations. The scripted calls poured in as well-funded and well-organized leftist groups targeted the committee. They wanted the Trump administration to get the same treatment they perceived Hillary Clinton had gotten. (I presume they weren't requesting that Trump be exonerated in a faux investigation with a predetermined outcome in which targets of the investigation get to pose as his lawyers and all interviews are off the record—a standard the DOJ reserves only for Hillary Clinton.)

Alas, the improbable election of Donald J. Trump was not probable cause for an investigation. Unlike the Clinton investigation, where we had a report from the inspector general for the intelligence community indicating there was evidence of classified information being housed in a nonclassified setting, no such evidence or investigation existed in the case of then President-elect Trump.

Before any congressional investigation could take place, the executive branch needed to investigate. Congressional committees are not law enforcement. They are oversight. The constitutional responsibility to enforce the law rightly belongs to the executive branch. We typically come in behind to oversee the executive branch actions or inactions.

The DOJ has 110,000 people to enforce the law. The House Oversight Committee has sixty to conduct oversight. We were not equipped with the resources to be the tip of the spear on potential international manipulation. Nor should we be. We can make sure the investigation is thorough, complete, professional, and unbiased. But at that point in December 2016–January 2017, even before the inauguration, there was no investigation to oversee. There was no evidence to weigh. On the other hand, in the Hillary Clinton case there was already evidence. We had probable cause. Hillary Clinton had used her personal server to bypass the Federal Records Act. We knew which laws had been broken.

Furthermore, the classified nature of much of the information involved in an investigation of a foreign state actor precluded my committee from taking the lead on the collusion allegations. Only the House Intelligence Committee had the necessary security clearances to do a thorough investigation of executive branch activities.

Since I left Congress, both the executive branch and the House Intelligence Committee have investigated the Russia collusion allegations. A special counsel has been authorized. Most significant, a host of investigators have been exposed for inappropriate bias.

In a March 8, 2018, letter from House Oversight Committee chairman Trey Gowdy and House Judiciary Committee chairman Bob Goodlatte, the two cite the following among their reasons for requesting a second special counsel: "There is evidence of bias, trending toward animus," they wrote, "among those charged with investigating serious cases. There is evidence political opposition research was used in court filings. There is evidence this political opposition research was neither vetted before it was used nor fully revealed to the relevant tribunal."

As this book goes to print, we are learning even more from Inspector General Michael Horowitz about the bias and animus toward Donald Trump by the DOJ. His report provides irrefutable evidence of that bias, even though explicit admissions of guilt were not found in the records kept on official devices.

The selective use of facts, deliberate withholding of relevant evidence, and toleration of politically motivated "unmaskings" of private conversations of Trump campaign officials have come to define the investigations of the Trump administration.

The most obvious example of the biases with the intelligence community is available in the voluminous record of more than fifty thousand text messages exchanged between FBI special agent Peter Strzok and FBI lawyer Lisa Page around the time of Trump's election. The texts are a rare window into the personal conversations of members of the intelligence community.

Before the election, the two are unambiguous in their support for Clinton, with Strzok writing, "God Hillary should win. 100,000,000–0," and Page responding of Trump, "The man cannot be president." Even more telling is the exchange revealed in the June 2018 OIG Report. We already knew of Page's text to Strzok: "[Trump's] never going to be president, right? Right?" The OIG, after extensive effort, recovered Strzok's telling response to that text message: "No. No he won't. We'll stop it." Who is we? We is the Deep State.

Curiously, an August 15, 2016, text from Page refers to a mysterious "insurance policy" in the unlikely event Trump is elected. The message, which many believe is a reference to fired FBI deputy director Andrew McCabe, reads, "I want to believe the path you threw out for consideration in Andy's office—that there's no way he gets elected—but I'm afraid we can't take that risk. It's like an insurance policy in the unlikely event you die before you're 40."

In the month before the election, the two exchanged a series of texts indicating they may have leaked information to major media publications. Shortly after the election, none other than Strzok himself was tapped to be part of the Mueller investigation into the Trump campaign. His text messages indicated he was reluctant to take the appointment because his gut told him there was "no big there there." In his investigative capacity, he was involved in interviewing National Security Advisor Michael Flynn.

Although Strzok was later removed from the Mueller investigation, later texts between the two show them discussing how to evade federal message-archiving requirements in their communications.

While this one example may not reflect the views of everyone in the intelligence community, many of whom are dedicated and patriotic, the volume of leaking indicates Strzok and Page were not alone in their desire to undermine this president.

Since the election of Donald Trump, the intelligence community has become a lot less covert in its attempts to sabotage the president. We have seen leaks, both classified and unclassified, at the highest levels. Everything from the president's conversation with the prime minister of Australia to telephone conversations between Flynn and Russian ambassador Sergey Kislyak has been released to the media by high-ranking Deep State operatives.

The intelligence community was not nearly so forthcoming when Hillary Clinton put our nation's secrets at risk by illegally housing them on an unsecured server located in her New York home. The rules were different for Obama administration officials.

Democratic Double Standards

In addition to the energetic vigor with which leaks and investigations unfavorable to the new administration were pursued, there were other changes I noticed in political, administrative, and legislative interactions.

After Trump was elected, I regularly got peppered with questions from the media about whether I was working with the White House. Always there was this implication of sinister wrongdoing if I spoke to anyone there.

Prior to the 2016 election, Oversight Committee ranking member Elijah Cummings gave me the impression he was regularly talking to advisor Valerie Jarrett at the Obama White House.

When Cummings and I were united in the need for changes,

I would leave it to him. He would coordinate directly with the White House. He was the Democrat! It was no problem. That happened on the Secret Service, problems with the Chemical Safety Board, the DEA investigation, and many more.

I never saw it directly, but I got the sense he was picking up the phone and calling them routinely, if not daily. I don't think there was necessarily anything wrong with the coordination. He would get back to me and say, "They're going to take care of it," or "They understand." It was just a given that he was coordinating directly with the White House.

Sometimes during those years I would call Neil Eggleston, White House general counsel, and talk some things through. But that wasn't happening week in and week out, like Cummings's communication appeared to be. It was benign.

But when President Trump took office, the rules changed. The media and the Democrats viewed with suspicion any coordination between congressional Republicans and his appointees. What had been commonplace for the Obama administration was now depicted as a form of corruption in the Trump administration.

An obvious double standard developed in the expectations of the relationship between the president and his attorney general. Democrats and media had no problem with President Obama overseeing the DOJ. The tight relationship between Eric Holder and President Obama had been unquestioned and well known. Loretta Lynch was viewed as nothing if not a loyal soldier to President Obama. Not so with President Trump and Attorney General Sessions.

Today the media would have you believe the Department of Justice functions independently from the president—that any coordination is somehow obstruction of justice. Media reports cited Trump's efforts to stop the recusal of AG Sessions from the Russia investigation as grounds for obstruction. Other reports cite Trump's dismissal of FBI director James Comey as obstruction.

In a classic case of projection, Democrats unwittingly exposed this double standard in an exchange during the Senate Judiciary Committee's June 18, 2018, questioning of DOJ inspector general Horowitz. Senator Kamala Harris (D-CA) asked Horowitz who else was shown the damning OIG report of the Clinton email investigation in advance. "Did you provide it to the White House?" He responded they had not. But then he added, "We've had instances, as I could testify to in terms of, for example, Fast and Furious, there was disclosures to the White House about it for various reasons that can be, there can be a basis to do it. . . . I think it is something you would have to direct the department for an answer."

What if the Trump White House had been shown the OIG report ahead of time as the Obama White House was during Fast and Furious? Per the Constitution, Donald J. Trump is the leader of the executive branch, including the DOJ. He is the president of the United States, and he is rightfully the one in charge of the DOJ.

While Trump is accused of obstruction for managing the DOJ, President Obama freely commented on open investigations without mainstream news outlets mentioning obstruction. That happened in the IRS case when Obama, in an interview with Bill O'Reilly on Super Bowl Sunday, was asked about the IRS scandal.

The president said there wasn't even a smidgen of corruption, yet there were five open investigations! House Oversight, House Ways and Means, Senate Finance, Senate Judiciary, and the inspector general all had open investigations, and none had been completed. President Obama had the gall to say there wasn't "even a smidgen of corruption."

He reached a similar conclusion in the Hillary Clinton scandal. He told Fox News' Chris Wallace in October 2016, "I can tell you that this is not a situation in which national security was in danger." "Here's what I know. She would never intentionally put America in any kind of jeopardy . . . she has not jeopardized America's national security." How did he know that? Few outside conservative media even questioned it.

Even James Comey later admitted that the comments to Fox News, 60 Minutes, and others were inappropriate.

Now President Trump is asked direct questions about his campaign, and the Democrats are all too excited to yell "obstruction of justice!" And the national media is willing to write stories about obstruction of justice at every whim by a Democrat. They never wrote these stories about President Obama's interference with Justice Department investigations. Some of my former colleagues went so far as to suggest he should be impeached for the very things President Obama had done on a regular basis.

Ungoverned and Ungovernable

The rules by which cabinet secretaries must live have also changed. Not only is the confirmation process longer, but once the appointees take office, they are often undermined by their staffs.

The Senate has confirmed President Trump's nominees at a snail's pace, particularly when compared to the rate at which Senate Republicans confirmed Obama nominees eight years earlier. Fifteen months into the Trump presidency, 40 percent of his nominees still awaited action. Although a recent rule change eliminated the minority party's ability to block a confirmation vote, Senate Democrats have used the cloture rule to delay confirmation votes at least eighty-nine times. Cloture votes require thirty hours of debate per nominee. Senate Republicans used that procedure on confirmation votes only eleven times during the eight years Obama was in office. These tactics are being used even when the nominee has stellar credentials, such as the case of Richard Grenell. Trump nominated Grenell to be ambassador to Germany in September 2017. Democrats successfully blocked his confirmation until April 2018 despite Grenell's valuable United Nations experience and foreign policy background.

The May 2018 confirmation hearing of CIA nominee Gina Haspel demonstrated the double standard applied to Trump nominees. While Senate Democrats voted convincingly to support Haspel's predecessor John Brennan, who had a history of limited engagement in enhanced interrogation tactics, they withheld their votes from Haspel over the same concern. Republican senator Tom Cotton rightly called out the hypocrisy at her confirmation hearing, later telling Fox News, "As far as I can tell the only Democrats who oppose Gina Haspel are in the Senate. I guess it's because they are blinded by their insane hatred for President Trump."

The hypocrisy and double standards do not end when the nominee is confirmed. Once Trump appointees take their place in the federal government, they face levels of scrutiny and sabotage not seen in at least eight years.

A relatively new cabinet secretary in the Trump administration probably had the best explanation. He told me: "Imagine running a political campaign and you win. Now the bad news is, you have to actually govern, implement, but the people they give you to work with are all of your opponents' people. They are the people that are supposed to implement the new direction and the new strategy. You can't turn the ship. You can't fire them. You can't demote them. You can't hire new people. Consequently you can only do a fraction of what you were sent there to do."

Gina McCarthy, EPA administrator, could literally be fooled into believing EPA's highest-paid employee was an undercover CIA operative. Trump's EPA secretary, Scott Pruitt, gets a story in nearly every major news outlet for flying first class. McCarthy's mistake didn't get a fraction of the coverage that then secretary of health and human services Tom Price got for taking charter flights.

In this case, the media is right to scrutinize cabinet secretaries and their spending. Pruitt certainly deserves every bit of scrutiny he has received. But that scrutiny should be equally applied. When Secretary of Housing and Urban Development Ben Carson spent some thirty-one thousand dollars to redecorate his office, it seemed every major news outlet wrote a story. But when Secretary Clinton spent billions to design fancier U.S. embassies abroad, mainstream coverage of the story was sparse. I know—because I investigated that travesty of a policy and pitched it to the media. CBS News and Fox News can be commended for their willingness to cover the story. But try searching both stories and see what comes up.

CHAPTER 12

The Wall: Why the Deep State Doesn't Want It Built

The chasm between President Obama's immigration policies and President Trump's is both wide and deep. Unfortunately, another chasm exists between what Trump has promised and what the Deep State has been willing to deliver—despite broad public support for the Trump agenda.

Top-heavy federal agencies still filled with Obama holdovers and advocates of big government have failed to implement President Trump's policies, despite support and resources from the administration. They and a coalition of groups who benefit from illegal immigration have little incentive to actually solve this problem. The result has been a missed opportunity—one that can be recovered when we finally build the Wall.

The Problem: Mismanagement or Subversion?

There is no excuse for the failure of federal agencies to actually implement President Trump's policies. Through a combination of continuing failed policies, resisting new directives, and mismanaging of resources, our federal agencies have done the American people a disservice.

I'm not talking about our brave border agents, too many of whom have made the ultimate sacrifice to protect America. Those men and women on the ground are not the Deep State. They are on the front lines and very much want to protect Americans from the criminal element that took daily advantage of our under-manned borders and permissive immigration policies before and during the Obama era.

Many of them, like Brandon Judd, president of the National Border Patrol Council, are frustrated with the failure of upper management to capitalize on the opportunities the Trump administration has provided for border enforcement. In testimony before the House Oversight Committee in April 2018, Judd told the committee:

> *While significant progress has been made in securing our border over the past year, much work remains to be done. President Trump has worked tirelessly over the past year to improve border security and stop illegal immigration. He's made it crystal clear that he intends to finally secure our Southern border with Mexico and fix our broken immigration system, but sadly career bureaucrats and Obama holdovers at DHS, CBP and ICE have slowed our progress. Whether it's the continued implementation of the catch and release policy or mismanaging manpower resources, CBP management continues to perform poorly. I implore the Members of this Subcommittee to use your oversight powers and jurisdiction to hold CBP management accountable.*

The Catch-and-Release Program

In 2013, President Obama reinstated a program of catch-and-release for illegal immigrants caught crossing the border. Under this protocol, the vast majority of illegal border crossers were simply released into the United States. This included, in too many cases, repeat offenders and felons previously convicted of violent crimes. In 2013, the Center for Immigration Studies reports, only 25 percent of deportable border crossers were actually charged with a crime. The rest were given a notice to appear before a judge and released. The *Washington Times* reports many Central Americans referred to the notices as "permisos," or free passes.

All of this was supposed to change under President Trump.

With the president's campaign rhetoric promising to deport illegal border crossers, illegal immigration fell steeply in the months after his inauguration. By April 2017, Trump declared that apprehensions at the Southwest border were down 61 percent. By July 2017, apprehensions were down 53 percent from the previous year.

But Judd says catch-and-release actually continued. "The word got back to all of these countries that hey, the catch-and-release hasn't ended," Judd told the *Washington Times*. "The people that are supposed to be enforcing President Trump's policy, they're just not. They're not following through on the promises he made."

As a result, illegal border crossings *increased* 203 percent between March 2017 and March 2018, creating a strain on the system that further exacerbates the problem. U.S. Immigration and Custom Enforcement is unable to free up beds. Prosecutors are unable to keep up with prosecutions amid the stream of illegal crossings. A Homeland Security press secretary admitted the agency is "severely constrained by litigation, court rulings, and debilitating legal loopholes that limit our ability to carry out our mission." In short, a de facto catch-and-release policy has taken hold.

In April 2018, the Department of Justice announced a zero-tolerance policy for prosecuting illegal border crossers. President Trump signed a proclamation sending the National Guard to the border in an effort to shore up manpower. Attorney General Jeff Sessions directed prosecutors to aggressively pursue cases, saying in a statement, "To those who wish to challenge the Trump administration's commitment to public safety, national security, and the rule of law, I warn you: illegally entering this country will not be rewarded, but will instead be met with the full prosecutorial powers of the Department of Justice." Prosecutions are an effective deterrent because once someone is charged with a misdemeanor, their second illegal entry can be charged as a felony. Once convicted of a felony, the chances for a person to immigrate legally drop dramatically and the penalties for the crime become more severe.

The addition of prosecutors, judges, Border Patrol agents, and even National Guard troops will only go so far if Obama holdovers who oppose those policies hold the power to implement them.

Manpower on the Border

The other major problem holding back President Trump's immigration agenda is a management issue. Once again, we aren't talking about the frontline people who are actually putting their own lives at risk to secure our borders, but middle- and upper-level managers who fail to take the action required to accomplish their mission.

Brandon Judd, who is a Border Patrol agent in addition to his responsibilities with the National Border Patrol Council, took his frustrations to Border Patrol chief Ronald Vitiello in an April 2018 letter:

"President Trump promised we would see an increase in the number of agents on the border, but as of today, that promise hasn't materialized. Not just in hiring, but also in the proper deployment of very limited resources," Judd said. The letter reveals some shocking numbers.

In one eye-popping example, Judd described Border Patrol assignments at the McAllen, Texas, station. With a total of 700 agents assigned to the station, there are some 400 on duty on any given day. One Sunday in March 2018, Judd pointed out that only 50 of those 400 were actually patrolling the border. That's 12.5 percent. The rest were assigned administrative or management tasks—essentially sitting behind desks. He said there was a point during that day when 16 agents were responsible for patrolling 55 miles of border—with five of those agents on the waters of the Rio Grande.

Although Trump has attempted to facilitate the hiring of more Border Patrol agents, the hiring process is lengthy and contains multiple bureaucratic hurdles. In November 2017 the GAO reported the agency's average hiring rate was a measly 523 agents a year. Meanwhile, 900 a year leave—either to retire, resign, or transfer elsewhere. Even with the call for more agents, the total number has dropped from 21,444 agents in 2011 to 19,437 in 2017, leaving the Border Patrol two thousand agents short of what they can fund.

Activating the National Guard may help. Arizona congresswoman Martha McSally is supportive of the move but acknowledges the management issues. "Right now, we need both things to happen," she told the *Washington Times*. "We need the National

Guard to be deployed; I fully support it. . . . But the mid-level management needs to take a fresh look at how we're deploying the agents we have. So we need to do both."

Frustration with the pushback from midlevel managers is growing. In a November 2017 letter to President Trump, Chris Crane, president of the National ICE Council, representing deportation officers, wrote:

"While officers view the President's position on enforcement as courageous, the Trump administration has left all of the Obama managers and leadership in place, a group that ICE Officers know after the last eight years to be completely incompetent, corrupt and anti-enforcement." He added, "Our corrupt and grossly incompetent managers protect one another and cover up for their own misconduct, demoralizing our entire workforce." Crane went on to say that "any law enforcement officer or soldier will tell you that if their organization and leadership are dysfunctional and unsupportive of their mission, even the best policy from higher ups won't result in mission accomplishment."

These holdovers from the Obama administration are not just thwarting the president. They are putting Americans in danger—something the last president did with impunity.

Drug smuggling and human trafficking across our borders is a real thing, and a dangerous burden not only for our country as a whole but particularly for the border states of Texas, Arizona, New Mexico, and California. Sadly, so is criminality. Estimates of how many immigrants are criminal vary widely. The Migration Policy Institute says about 820,000 immigrants have criminal convictions, and ICE says 1.9 million people are criminals. Incarceration rates are even more difficult to find. The *Texas Tribune*, in 2016, quoting ICE, noted that nearly 5 percent of all inmates in Texas prisons are illegal aliens.

Here is what we do know for sure, again going back to ICE statistics for 2015. Illegals had 12,307 convictions for DUI; 7,896 convictions for dangerous drugs; 1,963 convictions for burglary; 1,347 convictions for domestic violence assault; 101 homicides; and 216 kidnappings.

In April 2016, my House committee held a hearing on criminal aliens who had been released by the Department of Homeland Security. Here is part of what I said at that hearing:

> In a three-year period, Immigration and Customs Enforcement has released more than 86,000 criminal aliens into the American public. These are people that were here illegally, got caught committing a crime, were convicted of that crime, and then instead of deporting them, they were just released back out into the United States of America. All told, they had more than 231,000 crimes that they were convicted of, 86,000 of these people.
>
> This administration's failure to secure our border, enforce immigration laws, and hold criminal aliens accountable creates an ongoing threat to our public safety and sometimes delays consequences for innocent Americans. And many of those losses are preventable. The numbers became real in February of 2015 in a National Security Subcommittee hearing. During the hearing, we heard testimony from Jamie Shaw, whose 17-year-old son was murdered by Pedro Espinoza, an alien living in the United States illegally. Mr. Espinoza had been released from jail on a conviction for brandishing a weapon before the Shaw slaying. This is a weapons conviction.
>
> We also heard from Mike Ronnebeck, the uncle of Grant Ronnebeck. Grant was 21 years old when he was killed in

Mesa, Arizona, while working an overnight shift at a local convenience store. The guy is just working at the convenience store late at night trying to do the right thing. The alleged killer was in removal proceedings due to a burglary conviction but released by ICE on a $10,000 bond, and Grant was killed.

The Ronnebeck and Shaw families are not the only victims of crimes committed by aliens unlawfully present in the United States. Today, we continue to put names and faces with individuals whose lives were changed forever by the death of a family member killed by a convicted criminal alien. The common thread among these stories you are hearing today is that each of them were preventable. If ICE had only followed the law, it is highly likely that these witnesses would not be sitting here today grieving the loss of another loved one.

And I thank the family members that will be joining us on the second panel. They are heart-wrenching stories, and it was preventable. It didn't have to happen. You could have deported them and you chose not to, and it is just infuriating.

President Trump's election is changing things. The president signed an executive order January 25, 2017, that expanded ICE's enforcement focus to include removable aliens who have been convicted of any criminal offense, have been charged with any criminal offense that has not been resolved, have committed acts that constitute a chargeable criminal offense, or have engaged in fraud or willful misrepresentation in connection with any official matter before a government agency.

That last criteria—fraud—addresses in part the number of immigrants asking for asylum because they face persecution for race, religion, or even political opinion in their home countries. Somehow the number of those granted asylum in the United States jumped from 13,931 in 2012 to 36,026 in 2013. Most of those were from El Salvador, Guatemala, Honduras, Mexico, Ecuador, and India. How many of those claims might be fraudulent? Both AARP and *International Living* magazine rank Ecuador the fourth-best country in the world in which to retire!

Overall, some 105,736 criminal aliens were arrested in 2017, a 12 percent increase over 2016.

So, when we look at facts, and leave political passions, fake news, and liberals' false accusations behind, it is obvious the kind of damage that the so-called sanctuary cities movement is doing to our country.

The term *sanctuary* itself carries no legal meaning, and it varies by jurisdiction. In essence, these are actions by cities, counties, or states that limit how much local law enforcement can cooperate with federal immigration officials. San Francisco, for example, prohibits city employees from assisting ICE. Right now, legislatures in five states—California, Oregon, Connecticut, Rhode Island, and Vermont—have voted to become "sanctuary states," although some towns and cities within those states have voted to "opt out."

To be polite, the sanctuary city "movement" is misguided and dangerous. More than three hundred jurisdictions are now self-designated "sanctuaries." Deliberately blocking and prohibiting communication between local law enforcement and federal immigration agencies is ridiculous. At a House Judiciary Committee hearing in February 2018, Congress heard from the president of the Denver Police Protective Association, a Stanford University School of Medicine psychiatrist, and a Texas sheriff. "These sanctuary cities are a cog in the expanding opioid crisis," Texas sheriff A. J. Louderback told the committee.

This is not anti-immigrant rhetoric. The DEA reports that 80 percent of the illegal opioids sold in this country are brought in by foreign criminal organizations, primarily the Mexico-based drug cartels and specifically Sinaloa.

Sanctuary cities simply limit the ability of law enforcement to do its job. In places not burdened like this, laws are enforced! In February 2018, ICE led an operation in Oklahoma in which ten people were arrested for heroin trafficking and possession of drug proceeds. According to ICE, the targets of the operation supplied two-thirds of the heroin in Tulsa. Of those ten suspects, six were illegal aliens and two others were arrested for immigration violations.

The Department of Justice is suing California to allow federal enforcement within its borders. The Department of Homeland Security is publicizing which jurisdictions are blocking cooperation with federal authorities. President Trump is cutting off federal grant funds to those same jurisdictions. He also has dispatched the National Guard to help guard our border against protest caravans of illegals.

How can anyone dispute that the problem is real?

Why the Deep State Hates the Wall

Back in Washington, the unfettered immigration agenda is driven by big business, certain big industries, and especially by the Democratic Party—but carried out by unaccountable bureaucrats who also benefit from the chaos of rampant illegal immigration.

It's a useful case study in the ways Trump's agenda is being thwarted.

Why is this idea of border security so controversial? Shouldn't it matter to everyone? It should. It should matter to conservatives and moderates and those of us who care about the very idea of America, our unique values and place in the world, and America's exceptional history and future. Why would anyone push for blanket amnesty for those whose very first act in this country was to break the law?

The Deep State benefits from illegal immigration. As long as immigration is "an issue"—and as long as it continues, we need more and more government to figure it out. We need to do more studies, hire more consultants . . . for more decades. The Deep State benefits from the chaos that results in demands for ever more government. They also benefit from the election of politicians who grow government. Which brings us to the Democrats, who depend on escalating unresolved immigration issues to stir up anger and motivate immigrant voters both legal and illegal.

Big business obviously benefits from artificially low wages that allow them to profit. And big-business cronyism goes hand in hand with the big-government Deep State.

Illegal Immigration Is Very Profitable

Some of the Deep State resistance to border enforcement results from ties to entrenched corporate interests. Let's look at the technology industry, headquartered in California's deep blue Silicon Valley. How does unfettered immigration benefit them?

For years, big technology companies have argued that hiring foreign workers through the H1B visa process is critical to growth and success because there is a shortage of skilled American workers. Google, Amazon, and Apple have argued that those sixty-five thousand H1B visas are necessary. The H1B visa has become the poster child for "necessary legal immigration" in technology.

Except that is not the whole story. Many of those companies

that hire these workers are called outsourcing companies. It is those firms that do the actual hiring. The three largest H1B employers in 2012, for example, were Infosys, Wipro, and Tata, all based in India. Google, Amazon, and the big tech firms "rent" these workers, as do Disney and the University of California.

Surely it couldn't be that companies or even governments were using the H1B visa to pay lesser wages?

Surprise, surprise. The statistics show that the median wages paid to H1B visa holders are below what American citizens earn. And there are additional insults one can add as well. According to the *New York Times*, Craig Diangelo was among two hundred employees laid off from the Connecticut company then called Northeast Utilities in 2014. Diangelo had worked in the company's IT department for eleven years. He had been earning $130,000 a year. He was told he needed to train his replacement, an Indian H1B worker recruited from an outsourcing firm and making $30,000. If he refused to train his replacement, he would not get his severance.

"The problem is that my job is still there," he told the *Times*. "I went away. The American workers went away."

Plus there is another benefit to big tech. The H1B visa is not transferable to another employer. So the company gets an immobile worker. An American worker could change jobs or move away for a better opportunity. The foreign worker is stuck.

Some people have called the system indentured servitude.

And while the definition of a H1B holder is supposed to be a person with specialized or unique talents, that is hardly so. Back in 2007, the New York City Department of Education (DOE) wanted to hire more teachers, particularly in troubled schools. So did they advertise all over the United States for young talented teachers who wanted to move to New York City for a full-time job? Of course not. They held a regional hiring fair in Vienna, Austria. Some people who barely spoke English came to the fair from Slovakia, Hungary, and Eastern Europe. They were hired! The practice continues today. The New York City DOE filed for 192 labor condition applications for H1B visa holders from 2015 to 2017. In 2018, the top H1B visa employer in Brooklyn was JPMorgan Chase Bank. The city's department of education was second.

And how about the idea that H1B visa jobs are only for supremely talented unique positions? Nope. The top fifty occupations for H1Bs in 2018, according to myvisajobs.com, included physical therapists, graphic designers, Web developers, and marketing managers. The list even included public relations specialists—one of the most popular majors at American colleges and universities.

Let me tell you who else benefits from unfettered immigration: the hospitality industry, which employs 15.2 million people in restaurants and hotels in the United States. A 2008 Pew research report estimated that 10 percent of those workers, or 1.5 million people, are here illegally. Obama supporter and TV restaurateur the late Anthony Bourdain attacked President Trump and declared that if illegal immigrants were deported, "Every restaurant in America would shut down."

No, every restaurant would not shut down. The only ones that would shut down would be those that refuse to hire American workers and pay them the market wage. Prime example in the restaurant industry: President Obama banned tip pooling in restaurants. That practice means tips are shared by servers *and* kitchen workers, many of whom are immigrants. It was effectively a policy that incentivized the use of illegal immigrants because kitchen wages not supplemented by tips are too low to attract legal workers.

Already under President Trump, the Department of Labor has rolled back part of that law along with other burdensome unfair regulations that have driven so many restaurants to hire illegal workers in the first place. As of March 23, 2018, some aspects of tip pooling are allowed again.

Is it so controversial to suggest that American citizens should have priority access to American jobs? President Trump's "Buy American, Hire American" executive order shouldn't be controversial. Let us not forget that Trump himself has spent a career in the hospitality industry. He knows the industry can survive with a legal workforce.

The Impact of Misaligned Rewards and Penalties

I've often said that in public policy, you get more of what you reward and less of what you penalize. Immigration policy is a prime example. When we reward people for breaking the law with benefits that are not available to those who come here legally, we send a strong message. No one ever suggests that the children of those on the wait list to come here legally should get in-state tuition or automatic citizenship. But had those same parents come here illegally instead of waiting their turn, all kinds of groups would be fighting for them. The incentives are all wrong.

Under our system, we send the message that law-abiding people need not apply. We only want people who are willing to break the law. But is that really what we want? Not according to the results of the 2016 presidential election. It's not what President Trump wants. Nor the people who voted for him. The fact that some industries, some federal employees, and one particular political party benefit from this permanent underclass of immigrants should never justify jeopardizing the safety and security of the rest of America. By releasing asylum-seeking families who illegally crossed our borders and giving them legal work permits, the Obama administration actually incentivized traffickers to use children unrelated to them to pose as families and qualify for the free pass into the United States. We got more of what the policy rewarded—fraudulent asylum claims.

We so often hear, as I've mentioned, that we are a nation of immigrants. And that is absolutely true. We also hear many business leaders, politicians, and entertainers recount their family histories. Apple founder Steve Jobs's father was a Syrian immigrant. Facebook founder Mark Zuckerberg's wife's parents were refugees from China and Vietnam. Google's chief executive, Sundar Pichai, is from India. A father born in Turkey raised Muhtar Kent, the chief executive of the Coca-Cola Company.

Yes, all true. For hundreds of years immigrants have come to America to make a better life for themselves and their children. My own great-grandfather, Joseph Carroll Chaffetz, came from Lithuania, a refugee from famine and a cruel and repressive Russia regime. His son, my grandfather Maxwell Chaffetz, was born in the United States and went on to become an FBI agent. He gained some fame when he was involved in the capture of famed gangsters John Dillinger and Baby Face Nelson.

I am a huge proponent of fixing legal, lawful, regulated immigration. But we have to face the sad fact that illegal immigration today, especially from Mexico, facilitates massive criminal enterprises on both sides of the border. In too many cases, it is not the kind of immigration that brought so many of our relatives here from places like Ireland or Italy or Russia. Our lax immigration laws have enabled drug cartels, human traffickers, paid coyotes, and would-be terrorists from around the world to exploit our compassion and endanger our communities. We want to attract the ambitious achievers of past generations. We want to draw the best and brightest seeking to work hard for better lives. For those people, we must continue to make America a welcoming place—a place governed by fairness and the rule of law.

Here is the bottom line: the Deep State feeds on overregulation, needless government, and entwined private and government interests and contracts. We need a real solution. We need the Wall. Not a puny fence. Not a passable river or a walking path with guardrails that is ADA compliant. A wall. Our borders will not be safe until the Wall is built.

CHAPTER 13

Taming the Deep State

Even before the 2016 election, it was obvious to me that Washington, D.C., and the Deep State that runs it were only going to be fixed by bold disruption. We finally have a president who is willing to do that. He can do a lot—more than anyone anticipated, actually. But as we clearly see, the Deep State is waging all-out war against this president. The president needs Congress, both the House and Senate, to enact policies to restore our country.

Unless we can find a better way to check the power of those who enforce the law, they become a law unto themselves, practically exempt from following the very rule of law they are charged with enforcing.

Let's explore three categories of reforms, all of which must be addressed if the Deep State is to be checked. We must start by empowering Congress to use, and perhaps modernize, existing constitutional checks against the executive branch that have fallen into disuse.

We have to restore the penalties, rewards, and incentives that govern the federal workforce. And finally, we must expose corruption by bringing hidden things into the light.

It all starts with asserting the power of inherent contempt.

Giving the Inherent Contempt Power Teeth

The power of Congress to compel the production of information from the executive branch is undisputed. The Supreme Court has repeatedly upheld that power, as well as the authority to impose fines or even the more draconian measure of jail time. But I found today's Congress reluctant to rock that boat.

The requirement that Congress rely on the U.S. district attorney for the District of Columbia to enforce compliance to its subpoenas must be abandoned. Unfortunately, partisan politics makes a simple fix complicated. Over the years, both parties have bemoaned the lack of cooperation by the Department of Justice, but never at the same time.

Democrats wanted the power to enforce subpoenas against Bush administration officials but lost their passion when Barack Obama was elected president. Now that Donald Trump is president, the shoe is once again on the other foot. Fortunately, there is hope.

A range of options exists for restoring to Congress the ability to enforce contempt citations. Legislation could be passed denying the district attorney the ability to substitute his own discretion for that of Congress. Congress could unilaterally impose fines, even going so far as to withhold the fines from federal paychecks. As a last resort, Congress could even return to the days of remanding into custody those unwilling to cooperate.

Having failed to assert enforcement powers in the past, Congress must take some of the blame for the emergence of today's powerful Deep State, which now routinely refuses to cooperate with Congress. While I saw little evidence of congressional action to restore its own enforcement power during my time in Congress, Trey Gowdy offered some hope in a June 2018 interview on *Fox News Sunday*. Speaking of Justice Department stonewalling of congressional subpoenas, Gowdy told Fox News' Chris Wallace, "We're going to get compliance or the House of Representatives is going to use its full arsenal of constitutional weapons to gain compliance." We can only hope. There has also been movement to expedite a path for judicial access by Congress. While I prefer to see Congress assert its own power instead of relying on that of the court, this option is still better than the status quo.

Judicial Access

There's a reason the organization Judicial Watch is more successful at getting information than congressional oversight committees. They have a direct route to the courts! For the judicial path to be effective, Congress would need an expedited pathway to the courts for immediate consideration. Administration officials are afraid of courts because they incarcerate and impose fines—something they need not fear from Congress.

To see why expedited access is required, we need only look to the languishing Fast and Furious court case.

If we wanted to get information that agencies were withholding from us, we had to hold them in contempt, sue them in federal court, litigate whether we belong in federal court, argue whether the case is ripe for the judge, and wait for the judge to rule on those questions. Then and only then could the judge get to the merits. That takes two to three years, if not more.

Agencies know that even when Congress has a right to information, it pays for the agencies to withhold it. Even if they cite a bogus excuse, they know it's going to take years. In that time, committee leadership will change, public appetite for the information will ebb, and new scandals will come along to replace the old ones.

In addition, going to the courts requires us to take these things to the House floor and get the whole body to authorize a lawsuit. That, too, would have to change. Putting something on the floor, occupying every member's time, and spending hours debating whether we ought to get these documents is tough to do. Something like Fast and Furious, with big headlines, a murder, allegations of negligence—that is doable. You can get floor time and the attention when it's something like the IRS targeting matter. But what about when the head of the U.S. Chemical Safety Board lies to Congress? Are they going to want to use floor time for that? No. So he will get away with it. Which he did.

Not every oversight matter is something you can fairly ask leadership to point the whole U.S. House toward. It's just not feasible. The Deep State apparatus knows this. They can just withhold information and run out the clock. Congressional committees should be empowered and authorized to move this process forward without the need for the full body to debate and vote.

Mandate Federal Prosecution of Perjury

It's time for the Trump Justice Department to apply the rule of law uniformly against all who lie to investigators.

This is the missing piece in congressional investigations. If the Justice Department won't prosecute those who lie to Congress, there is no deterrent to doing so. Unfortunately, Attorney General Jeff Sessions has not demonstrated any willingness to strengthen oversight by punishing liars. Two egregious examples include the EPA's aforementioned Chemical Safety Board chairman Rafael Moure-Eraso and former secretary of state Hillary Clinton.

In March 2017, the Sessions Justice Department announced they would not seek charges against Moure-Eraso, who lied to the House Oversight Committee. The committee had forwarded a criminal referral on Moure-Eraso in July 2015.

In two separate Oversight Committee hearings, Moure-Eraso's testimony to the committee was contradicted by facts and documents. Criminal referrals from Congress are very rare. Bipartisan referrals are even more so.

Ranking member Elijah Cummings and I jointly requested the criminal referral due to egregious differences between Moure-Eraso's testimony and the evidence against him. This was as bipartisan as it gets! More than one year after our initial request, DOJ decided they were not going to prosecute him.

Similarly, neither the Obama nor the Trump Justice Department has shown a willingness to prosecute Hillary Clinton for obvious lies to congressional investigators.

Lie in an FBI interview? Boom! But I've never seen a consequence for lying to Congress. They just don't do it. That has to change.

Impeachment Power

Article II of the U.S. Constitution gives Congress power to impeach the president, vice president, and all "civil officers of the United States." This includes cabinet officers. But not since 1876 have we successfully used it against those who run our federal bureaucracies. That is a role and responsibility we have simply relinquished, even though it is embedded in the Constitution.

Congress needs to once again assert this tool. Doing so requires no new legislative authority. Congress already has the power to do it, but so far it has not had the will.

Not everyone confirmed by the United States Senate turns out to be competent or to have the integrity to do the job the people need them to do. If Congress cannot impose a single degree of accountability, what good are we?

The *New York Times* in 2007 called for the impeachment of Bush administration attorney general Alberto Gonzales to "defend the Constitutional order." They wrote, "The real question is whether Republicans and Democrats are prepared to defend the constitutional authority of Congress against the implicit claim of an administration that it can do what it pleases and, when called to account, send an attorney general of the United States to Capitol Hill to commit amnesia on its behalf."

For once, I find myself in agreement with the *New York Times*. The difference is, this should be true all the time, not just when Republicans are in power.

Power of the Purse

The power given to the U.S. House of Representatives to initiate the budget process purportedly also gives us grand powers to use budgetary authority as a means of gaining compliance.

While this sounds good in theory, it doesn't work in practice because the federal budget process broke down many years ago.

We did try to assert this authority in response to the IRS targeting scandal. After John Koskinen lied to Congress and allowed the destruction of evidence, we voted to cut the budget of the IRS by hundreds of millions of dollars.

But Congress doesn't have a scalpel—only an ax. We could cut the overall budget, but we couldn't decide for them who would feel the cuts. Instead of giving up their bonuses or finding efficiencies, they instead chose to cut customer service. That way they could pass the pain on to the voting public in the hope that the resulting outcry would put pressure on Congress to restore the funds. This wasn't fair to the millions of people who faced hours-long waits for service from the IRS.

If Congress could summon the will to fix its dysfunctional budget process, perhaps a powerful check and balance could be restored. All that is missing is the will.

Rescind De Facto Lawmaking Authority

A little-known but insidious usurpation of congressional power is the Office of Legal Counsel (OLC). This executive branch agency writes legal opinions that direct the prosecutorial priorities of the entire DOJ.

If they write a memo indicating marijuana prosecutions are no longer a priority, those prosecutions stop. They may still be the law of the land, but the executive branch has indicated an unwillingness to enforce them. This practice of looking the other way on specific categories of crimes is a de facto lawmaking authority that enables the Deep State to override the will of Congress and undermine the separation of powers.

Imagine this fact: *fewer than twenty people hold the keys to steering the enforcement apparatus of the entire federal government.* In this role, they can virtually create new laws overnight by simply reinterpreting existing law. This authority has become far too broad. Congress must act to rein it in.

The Obama administration used this power to unilaterally rewrite the long-standing 1961 Interstate Wire Act in December 2011. Virtually overnight, the restrictions against online gaming disappeared—without any public debate, public hearings, legislation, or semblance of a democratic process.

One day the Wire Act precluded online gaming. The next day the same words were interpreted to mean exactly the opposite.

This was a radical public policy departure imposed on every state in America whether it wanted it or not. Whether you support online gaming is not the issue. It is unequivocally giving a career lawyer in the Justice Department the authority to make laws. That goes against every structural protection erected by our Founding Fathers to impose separation of powers.

More important, Congress should set strict boundaries on OLC authority.

Addressing Rewards and Incentives in the Federal Workforce

We must restore the accountability that has been lost through a series of disastrous decisions. With the federal workforce, as with federal immigration policy, we get more of what we reward and less of what we punish. The inability to get rid of the bad apples, utilize appropriate incentives, and minimize corruption has created a monster that can no longer be contained without significant reforms.

Federal employees in general are very well taken care of. A 2017 CBO report studied compensation for the 2.2 million–strong federal workforce and determined that the federal government, on average, pays 17 percent more in total compensation than the private sector pays for similar positions. That disparity is higher the less educated a person is, but federal employees are almost impossible to fire. Once fired, they are often reinstated by the Merit Systems Protection Board. Unlike most Americans, they enjoy both a 401(k) plan and a federal pension in addition to Social Security and Medicare. They are represented by powerful tax-exempt unions whose representatives are often on the federal payroll while they work on behalf of the union.

On top of all of those advantages, they also enjoy the full support of the Democratic Party, to whom their unions donate lavishly and almost exclusively.

Make no mistake, federal workforce reforms are the heaviest lift. Not only is the federal workforce itself change resistant, but its powerful union, political, and media allies will align against even the most benign efforts to restore accountability. That said, Congress has shown a willingness to address these issues. Furthermore, there are many within the federal workforce who recognize the damage being done by the failure to hold people accountable. The most promising recent development is President Trump's May 2018 signing of three executive orders designed to make firing federal employees easier. The Heritage Foundation reported in a June 13, 2018, report that the reforms are surprisingly popular with rank-and-file government employees. In a Government Business Council flash poll, a slim majority of federal workers—51 percent—said they support the Trump policy.

Get Rid of the Bad Apples and
Empower Managers to Fire People

This seems like a no-brainer, but I encountered a shocking amount of resistance to the idea of actually firing people who do things wrong. Even more baffling was the resistance to rewarding merit. They were all for bonuses and pay increases, but the culture in the federal workforce—particularly among government employee unions—finds it "unfair" to reward people based on merit.

Firing federal employees is so laborious that Congress had to actually pass legislation to enable the Department of Veterans Affairs to fire people after a 2014 Phoenix, Arizona, scandal in which patients died waiting for VA care. (President Trump signed that bill in 2017). With the advent of collective bargaining and more complex protocols for due process in the 1960s, individual managers lost their autonomy to impose consequences on recalcitrant employees. The personnel decisions of managers can be appealed by the employee to an outside board. Although most cases ultimately favor the agency, the hassle of going through the process is a deterrent to taking action against bad apples in the federal workforce.

This can be approached any number of ways. A good start would be passing legislation to make permanent the solutions in President Trump's executive orders. Those changes include:

- Shorter performance improvement periods. President Trump has imposed a standardized improvement period of thirty days before an employee can be fired. It shouldn't take six months to know whether an employee is able to improve.
- Reining in official time. Although the executive order does not end the practice of paying employees their full-time wage to act on behalf of the union, the practice is now limited to 25 percent of an employee's work day.
- Limiting union grievances. This order removed the option for federal employees to have a union arbitrator hear their appeal for major disciplinary action—a path that has long been seen as favoring the employee in disputes. Now grievances must be heard by the Merit Systems Protection Board or a federal court.
- Ending expungements. Finally, the executive order puts an end to the practice of deleting records of disciplinary actions to resolve complaints.

Some other possible solutions remain:

- Expedite the removal process for those guilty of certain categories of crimes, which may include tax delinquency, sexual harassment, and committing perjury.
- Empower managers to adjudicate conflicts within the agency, not through the Merit Systems Protection Board.
- Make federal employees at-will employees, as some states have done.
- Spell out circumstances when criminal referrals are appropriate.

Honestly, I have been open to any and all creative proposals for expediting the removal of recalcitrant employees. But the forces working against such efforts are formidable.

What We Are Up Against

To understand what we're up against, my experience with federal employee tax delinquency is instructive.

When I was a freshman, I realized Democrats owned all the levers of power at that time. I was elected in 2008—the same time as President Barack Obama. That was a wave election for Democrats. So, in an effort to find common ground, I went back and looked through the pieces of legislation Obama had sponsored during the very brief time he was a U.S. senator. Guess what? I found one that I actually really liked. His S. 2519 was the Contracting and Tax Accountability Act. The bill prohibited the awarding of contracts or grants to entities with seriously delinquent tax debt.

I thought it was a good idea. But I tweaked it before I introduced a similar bill in the House. I took the same idea Democrats loved when Senator Obama introduced it and I applied it to the federal workforce.

Each year, there is a report that comes out from the IRS. It highlights the nearly 100,000 federal workers who owe nearly $1 billion ($962 million in 2008) in taxes. When retirees and military are included, there are more than 276,000 people who owe $3 billion. The IRS was able to fire federal employees who didn't pay their taxes, but no one else could. The IRS was embarrassed when statistics indicated the high number of IRS employees who did not pay their taxes. Congress altered the law so IRS employees could be fired for not paying their taxes. A few years later, they became the most compliant agency. Imagine that . . . change the consequences and the compliance changed dramatically.

On March 3, 2010, I introduced HR 4735, which would terminate the employment of current federal employees and prohibit the hiring of future federal employees who have a "seriously delinquent tax debt." That bill did not address contractors, as Senator Obama's had.

Federal employees are treated quite well by taxpayers:

- **Job Security:** Federal employees enjoy tremendous job security. Since 2000, executive branch civilian full-time equivalent (FTE) employment, excluding the U.S. Postal Service, has increased from 1.89 million to 2.18 million, or 15 percent. Private sector employment decreased 3 percent, from 110.2 million to 107.1 million. *Source: Bureau of Labor Statistics establishment data, table B-1, seasonally adjusted*
- **Wages:** From December 2007 to June 2009, the number of federal employees earning more than $100,000 increased

46 percent while the number of federal employees making more than $150,000 more than doubled. During the same time period, federal salaries increased 6.6 percent while private sector and state-local government employee salaries increased only 3.9 percent. *Source:* USA Today *based on data from Bureau of Labor Statistics and Office of Personnel Management, December 11, 2009*

- **Turnover Rate:** Federal employee turnover rate—including layoffs, discharges, and quits—is 60 percent lower than the private sector average. *Source: Bureau of Labor Statistics Job Openings and Labor Turnover Survey (JOLTS) data from 2000 to 2009*

If at First You Don't Succeed . . .

I tried again in the 112th Congress. I submitted an op-ed to the *Daily Caller* in an attempt to build support for the federal employee bill. A few months later, in January 2012, the IRS released a new Federal Employee/Retiree Delinquency Initiative (FERDI) report showing even higher rates of tax delinquency among federal employees. Now there were 98,000 federal employees who owed more than $1.034 billion in unpaid taxes from 2010—a 3 percent increase. And while the number of tax delinquent federal employees remained fairly constant, the amount owed increased 72 percent.

This time I introduced two bills—one for federal employees and a second that would apply to government contractors as Senator Obama's bill had. These bills were introduced simultaneously (numbered HR 828 and HR 829 that year). I was actually able to get a Democrat to cosponsor the contractor bill, my House Oversight colleague Representative Jackie Speier of California.

A GAO report showed federal contractors owed $5 billion in unpaid federal taxes. If you're a company competing for a government grant or contract and you don't pay taxes, guess what? You're probably going to be able to undercut your competition. It's not a fair fight!

There are a lot of ways to make a business profitable when you engage in fraud. But why should government grants and contracts go to companies that aren't paying taxes? If you have a history of doing this, then you should not be eligible to win a contract.

This time both bills were marked up in committee. The contractor bill passed unanimously. Not so for the federal employee bill. In markup, Democrats argued the federal employee bill was purely symbolic and unnecessary.

Firing federal employees is difficult, but I tried to craft the legislation to allow management to fire those who were tax delinquent. The bill said if you were working with the IRS and had a plan to repay the past-due taxes, you were not in jeopardy of losing your job. If you were contesting the obligations that the IRS thought you had, then again you weren't going to be subject to discipline and departure. This was narrowly tailored to those who just don't pay.

In response, Democrats argued that people need a job if they're going to pay taxes. I would respond, "But they're *not* paying their taxes!" In circles we went, around and around, time and time again, with Representative Steve Lynch of Massachusetts and Representative Lacy Clay of Missouri doing all they could to make the argument in favor of federal employees who were tax delinquents.

The final report from the committee markup on the bill concluded with a minority note from ranking member Cummings that read: "All federal workers should pay their taxes. The Committee's efforts and energy would be better spent, however, by focusing on measures to strengthen the federal civil service and improve the efficiency and effectiveness of the federal government rather than by making symbolic gestures intended to perpetuate a negative image of the federal workforce."

A month later, the bill was brought to the House floor, where it passed, 263–114. As often happened during Senate Majority Leader Harry Reid's tenure, the Senate refused to consider the bill and it never became law.

A New Strategy

As I prepared to introduce the federal employee tax delinquency bill for the third time, I prepared a new strategy and separated the discussion on the two bills. They would be considered separately this time.

The Democrats loved it. Everybody saw the wisdom. Everybody thought that was the smart vote. Of course! Why should we give contractors money if they aren't paying their taxes? It made great sense and flew through committee.

Then I took the exact same principles—bent over backward even more to help them out. No way, no how, were the Democrats going to support anything that would dismantle the Deep State. It was an incredible, vivid double standard.

Democrats supported the contractor bill (renumbered for the 113th Congress as HR 882) but objected to the federal employee bill (now HR 249).

Of course, with Republicans' pathetic messaging machine, nobody ever heard anything about it. It was such a classic example of a missed opportunity. There are too many Republicans who don't want to rock the boat with federal employees.

Put the Merit Back in the Merit System

The second part of restoring accountability is rewarding merit.

A 2016 GAO report found 99 percent of federal employees are rated "fully successful" or better. The ratings are meaningless when everyone is supposedly above average.

In business, you get rewarded for not only taking risks, but also producing. You have to actually produce—do things. There will be all these tired arguments that government is different than business. But there are still metrics you can put in place for good customer service.

At a minimum, we should ensure that those accused of serious wrongdoing are ineligible for bonuses. Again, this should be a no-brainer. But it doesn't work that way. The IRS's Lois Lerner was getting 25 percent of her salary in bonuses each year, including the years she was under investigation. She received $129,000 in bonuses between 2010 and 2013, according to the *Washington Free Beacon*. For what merit was she being rewarded? Showing up to work? Undermining conservative political groups?

The Heritage Foundation indicates that "[m]anagers must currently develop extensive Performance Improvement Plans (PIPs) for the employees to whom they do not award step increases. These PIPs are very time-consuming. Employees can also appeal managers' decisions to deny them a step increase through union grievances or the Merit Systems Protection Board, and ultimately through the court system. These prospects strongly encourage federal managers to give everyone a scheduled WIGI (within grade increase), irrespective of performance. Congress should not make giving everyone a raise the only sensible option for federal managers."

According to the Office of Personnel Management (OPM), in 2014 seven out of every ten federal employees got a cash bonus!

It cost taxpayers $1.2 billion. Your government is working that well!

A Better Model

The opposite is true on Capitol Hill, where a younger workforce spends a few years on the Hill and then departs for the real world.

I once read a management book called *It's Not the Big That Eat the Small . . . It's the Fast That Eat the Slow*. One of its basic premises is that you get in, serve, and then move on. I believe in that. I have lived that. It is one of the reasons I left Congress after eight and a half years.

Government service should be viewed not as a lifetime career but as something you can do with a portion of your life. We need to find a way to bring people in, allow them to serve, and allow them to continue with their lives. We should stop trying to find parity with the private sector. Private sector employers generally contribute between 3 and 5 percent of salary for retirement. The federal government's contribution is between 15 and 18 percent of salary.

Restructure Benefits to Stop Prioritizing Longevity

The federal workforce is very old. In large part that's because the benefits are so lucrative and they are structured to encourage people to stay at least twenty years. Where else can you get a pension, a 401(k), and post-retirement health-care benefits? People don't leave. They can't find anything in the private sector that matches what they get from U.S. taxpayers. There are virtually no companies left where you get the equivalent of a 401(k) and a retirement package on top of it.

As a side note, I would exclude the military from these changes, as I believe their heroism has earned them a different degree of compensation. What they do is more than a job.

Slow the Revolving Door

If you come in and work for the government, you can't just have a revolving door and go to work contracting with the government, exploiting your government network in order to negotiate bad deals for taxpayers. We need at least a two-year prohibition on this practice.

What happens all too often when people finally leave the federal workforce is they use their government contacts to get contracts or to work for the very vendors from whom they used to purchase. Turning contacts into contracts is a lucrative business for the employee—but not so beneficial for the taxpayer. This is not always in the best interests of the American people as decisions get made based on relationships rather than cost or quality.

Unfortunately, this is probably most prevalent at the Department of Defense. There is an unwritten rule that senior leadership within the military will be hired by defense contractors after they depart as long as they're friendly. Look at the major defense contractors and you will see person after person with military history. On one hand this makes sense, as they're in a specialized industry, but it does make you scratch your head and wonder if there isn't a prevalent "go along to get along" attitude.

Particularly with purchasing agents, I think this is problematic. I don't want to disparage an entire group of professionals, but it is an issue.

While we all can provide oversight of every transaction, the immediate need should be on the no-bid contracts. In these instances, where there is no competition, you really have to wonder how that came to be. If we want to focus on the most prevalent problem, this would be it.

Reduce Union Influence

If we really want to move the ball forward on government employee accountability, we have no choice but to reduce the influence of government employee unions—a process that has already made substantial progress.

In reality, these groups work against the interests of taxpayers and actually fight hard to prevent any kind of accountability. I have never yet encountered a union willing to work with me on getting rid of bad apples. These organizations care little about the public good. There is a place for them to function as professional associations. But we need to take our cues from the states, who have made great progress in reducing this perverse incentive to work against the public. Union dues should not be mandatory. This is an issue the U.S. Supreme Court got right in its 2018 *Janus* decision. We should limit what can be negotiated through collective bargaining. We absolutely must end the practice of "official time"—in which we allow federal employees to conduct union business on the clock. President Trump's executive order is a good start, but we must codify restrictions through legislation.

Union Influence in the VA Bill

The history of the VA bill is instructive. What happened in Phoenix was inexcusable. At least thirty-five veterans died waiting for care in what one VA whistleblower called "the worst example of VA health care in the United States—period." There was bipartisan outrage and both sides wanted to do something. Ultimately, the House and Senate each passed a bill. The bill represented exactly the types of reforms needed across the federal workforce.

It empowered the VA secretary to fire, suspend, or demote an employee with only fifteen days' notice. The bill still allowed employees to appeal to the Merit Systems Protection Board (MSPB), but in an expedited time frame. MSPB would then have 180 days to issue a decision, a much longer period than the forty-five-day timeline set up in the House bill. Employees still maintained the right to appeal those decisions to federal court. But unions were livid. Heaven forbid these reasonable ideas should potentially spread beyond the VA.

Even with bipartisan support, the bill was opposed by the relevant unions and by some Democrats who feared it would eat away at civil service protections and due process. So afraid are unions of accountability that they will oppose legislation to address the unnecessary deaths of veterans at the hands of their union members!

Union Impact in the Workplace

Countless times I met with federal managers who begged me to give them the tools to fire people. It was the same story all over the federal government. I remember one particular manager's story. He had a toxic employee who he felt was incompetent and had no business working in his agency. But the employee had passed the short probationary time and was now virtually impossible to fire. The manager tried disciplining the employee. This made the employee even more angry and caused him to continually attempt to poison the entire office—embroiling everyone in constant drama. The manager tried to fire the employee. But the Merit Systems Protection Board came through and forced him to reinstate the employee and provide back pay for a long absence. Now the employee was more toxic than ever. And the manager had no recourse. Meanwhile, the employee's interests were being looked after by another very highly-paid employee who was on leave from his job so he could work for his union—another case of official time. This employee was receiving his professional-level salary to spend his day protecting a toxic employee on behalf of the union.

Ultimately the manager could no longer discipline the toxic employee. He had no power to do so. Productivity suffered. The work culture became toxic. And the good people transferred elsewhere. This happens across the federal workforce.

Please, Work with Me

I was invited to come speak to the National Treasury Employees Union (NTEU) some years ago.

Representative Elijah Cummings was a hero to them. They loved the Maryland Democrat. They invited me to come speak, too. Rarely has a Republican taken them up on it. Going into my first year as chairman, I accepted their invitation. I came to a D.C. hotel. The room was packed full of NTEU members from around the country. They are one of the largest unions, representing 150,000 employees across thirty agencies. There were hundreds of people in that room, from all over the nation.

They were surprised I was there. I got a nice lukewarm welcome. I tried to share with them areas where we might have common ground. Then I pleaded with them, please, work with me. "Among you are some bad apples. And that small percentage of people spoil it for everyone else. We have to be able to get rid of them. If you'll work with me on that, I'll help you on other issues."

I got a nice, polite applause. Like every other government employee union I encountered during my time in Congress, they had no desire to do anything to allow government to ever fire any employee for almost any reason. Even if people were arrested, most unions didn't think that was sufficient cause to fire someone. Pornography, sexual harassment, none of these is enough to them. How much more clear could it be that they are working against the interests of the American taxpayer?

They will not work with us. They will fight tooth and nail. Our only option is to limit the power they wield. We need to stop subsidizing union activities. That's why it is so important to stop paying for "official time" where a $150,000-a-year air traffic controller gets paid his full salary to work full-time for his union without ever spending a minute helping to guide an airplane. Taxpayers should not be subsidizing the organizations whose goal is to essentially fleece taxpayers.

Empowering Inspectors General

One powerful tool that is underutilized is the Office of Inspector General (OIG). We have seen the role of the OIG play out over the summer with the June release of the damning report of the Clinton email investigation. That is the work of one office of the inspector general—the one assigned to the Justice Department. Each agency has an independent OIG charged with investigating wrongdoing within the agency. That office has access to information that journalists and congressional investigators sometimes have difficulty obtaining.

IGs are appointed by the president but confirmed by the Senate. Think of them as the internal forensic auditors, with powers to seize documents and review information no matter how sensitive.

They are independent operators void of political drama. They should be viewed as the honest brokers who will take an in-depth look at what's happening or not happening within a department or agency. But some of their powers are limited. Furthermore, the executive branch has acted to limit them further. Their role must expand, not contract.

For instance, if an employee leaves government, the IG is no longer empowered to continue to be able to investigate that person. Many federal employees accused of serious wrongdoing simply retire rather than face discipline. We've seen this again and again as claims of sexual harassment proliferate. The perpetrator simply retires and begins collecting his fat government pension.

I personally remember incidents in which an IG was getting close to finishing an investigation only to have that employee scribble a note saying, "I hereby resign." The IG had to walk away.

The IG should also enjoy broad and unfettered subpoena authority. In August 2014, forty-seven IGs sent a letter to the House Oversight Committee expressing the urgent concern that their access to documents and other important materials during investigations was being curtailed.

In July 2015, the OLC issued an opinion limiting the IGs' authority to access certain information, including grand jury materials. I want the IG to be able to see whatever they need to see to get to the bottom of their investigations. As we will discuss in a moment, the level of grand jury secrecy is something we need to reconsider, particularly in the case of OIG investigations.

Finally, when the IG does make a criminal referral to the DOJ, Justice needs to act on it.

The DOJ often believes the ultimate remedy for internal wrongdoing is for someone to lose his or her job rather than be prosecuted. They would leave it to the agency to mete out discipline rather than spending resources to prosecute a federal employee. The consequence is nobody ever gets prosecuted. That's not even something a federal employee has to worry about. The Deep State knows this. They can pretty much operate with impunity. Perhaps this is one reason we saw the FBI acting with such brazen contempt for the law as they worked to ensure a Clinton victory in 2016. Even in the most provable cases, it is rare that the DOJ will ever prosecute a federal employee. There are a few examples, but they are incredibly rare and particularly egregious.

The issue of OIG authority has come to the forefront this year. In the debate over whether the Trump administration should appoint a second special prosecutor to investigate political bias at the DOJ, oversight chairman Trey Gowdy has pointed out that these limits preclude the OIG from doing a comprehensive investigation. Instead, many called for a special prosecutor. This costly and extreme measure is the right call, but only because of the limits placed on the OIG. If we can find a way to address those limits, we can save the time and drama of a special prosecutor in the future.

There are seventy-three IGs that together employ roughly 13,500 people. This is in addition to the Government Accountability Office (GAO), which employs some three thousand people to perform audits on agency spending. We have the resources to hold federal employees accountable. We just need to give IGs the authority they need to go the distance.

Starve the Beast

Limited government needs to be a priority. Government has a tendency to want to grow and self-perpetuate. It isn't efficient. We spend more than $900 billion a year on social programs in this country, but the bulk of that pays for salaries and infrastructure. Little of that water ever gets to the end of the row. To the extent there are better and more efficient ways to get things done, we should seek those options. Government should be a last resort for only those things that can be accomplished in no other way.

Deep State Kryptonite: Sunlight

The last step in solving our Deep State problem is embracing transparency. Secrecy is a stock in trade of the Deep State. They maintain their power by virtue of their ability to cover their tracks in the many cases in which the public would surely disapprove. Consequently, the Deep State has a serious weakness.

Sunlight, in the form of transparency and accountability, is a powerful cleansing agent for both exposing and disincentivizing corruption.

In *Federalist 51*, James Madison listed "a dependence on the people" as the primary control of government. As Justice Louis Brandeis famously said, sunlight is the best disinfectant. For the Deep State, sunlight is kryptonite.

In the absence of market forces to constrain government, the threat of public exposure creates its own restraints. For this reason, the ultimate check and balance must be transparency. Exposure and oversight is a powerful antidote for corruption. Unfortunately, the cure depends on the ability to prove a case. That requires access to documentation—something the Deep State currently controls almost with impunity. The ability to cover up the truth is a powerful tool in their arsenal, one they've proven loath to relinquish.

Unearthing the Document Trail

How different would our history look if we really knew the truth? The Deep State does not want you to find out, which is why they have become adept at burying the paper trail they are legally required to keep. The 1966 landmark Freedom of Information Act requires mandatory disclosure of government documents but carves out nine exemptions under which documents can be withheld.

A 2015 investigation by the House Oversight Committee revealed a pattern of stonewalling and abuse by government agencies that worsened under the Obama administration. Agencies were routinely denying or ignoring legitimate requests for information. The backlog of outstanding requests doubled under President Obama. As of June 2015, seven and a half years into the Obama administration's tenure, 330,000 Freedom of Information Act requests had been denied. Records showed requests were being fulfilled just 30 percent of the time.

On this front, there is actually a small win to report. FOIA reforms passed by Congress and signed by President Obama in 2016 take aim at the Deep State weapons of time, information, and secrecy.

The bill strengthened the law requiring government disclosure of documents by codifying a presumption in favor of disclosure outwardly championed by Obama. The bill also purported to close certain loopholes used by government agencies to inappropriately withhold documents. In theory, the bill would reduce the backlog of delayed requests.

Unfortunately, this legislation was a watered-down version of a bill federal agencies and Obama administration officials covertly killed in 2014. Obama's previous public statements notwithstanding, a long-delayed response to a FOIA request from *Vice*

News reporter Jason Leopold eventually revealed the president's behind-the-scenes efforts to kill the legislation.

A 2015 investigation by the House Oversight Committee during my tenure revealed a pattern of stonewalling. An exhaustive two-day hearing on the matter unearthed a parade of eleven witnesses, including journalists and other frequent FOIA requesters, that demonstrated just how broken the FOIA process was. In many cases, documents were only produced after the drastic and unnecessary step of filing and pursuing a lawsuit. Yet the very next day, FOIA administrators painted a picture of a well-oiled machine, rating themselves 5 out of 5 on presumption of openness and effectiveness of response systems. As my jaw dropped listening to this testimony, I finally had enough and told one government witness, "You live in la-la land. That's the problem."

In a January 2016 report, we released the following key findings:

- The executive branch culture encouraged an unlawful presumption in favor of secrecy when responding to FOIA requests.
- The Obama administration was unaware that FOIA was systemically broken.
- Agencies created and followed FOIA policies that appeared to be designed to deter requesters from pursuing requests and created barriers to accessing records.
- FOIA requesters had good reason to mistrust even fair and earnest attempts by agencies to fulfill requests. In the words of one requester, "something is desperately wrong with the process."
- The Department of State had numerous open requests that were nearly a decade old, making them arguably the worst agency with respect to FOIA compliance.

Despite the reforms, FOIA laws remain woefully inadequate. Further FOIA reforms are the low-hanging fruit of the transparency issue. Last-minute amendments to the 2016 bill exempted intelligence agencies, arguably giving them even more discretion to withhold incriminating information. If 2017 taught us anything, it is that intelligence agencies are just as vulnerable to being politicized as any other agency, if not more so.

More protections are needed to ensure agencies do not misuse exemptions they claim to be valid. A judicial check and balance would enable an appeal to the courts for a ruling on the application of exemptions.

Address Classification Abuse

The ability to classify documents is a critical tool to protect national security. It is also a perfect way to hide information that is embarrassing or that documents wrongdoing. Although I can't reveal classified information, I can confirm I've personally seen classified documents that do not meet any rational criteria for defending national security but do expose incompetent or nefarious behavior. The only reason they were classified was to protect people from embarrassment or to thwart the ability of Congress to oversee the bureaucracy.

When FBI director James Comey famously notified me that the Hillary Clinton investigation had been reopened before the presidential election, the letter he sent was not classified.

Ironically, it was Clinton's predecessor in the United States Senate who literally wrote the book on overclassification. The late senator Daniel Patrick Moynihan's solutions are as relevant today as they were twenty-five years ago. He established the 1994 Commission on Protecting and Reducing Government Secrecy to address the classification challenges. The time has come to revisit this concept. Bipartisan support exists for reforming the classification process.

Moynihan argued that when everything is classified, nothing is classified. I believe we are handing out too many security clearances, classifying too many documents inappropriately, and being insufficiently selective in our vetting. Most important, we need to come to a bipartisan agreement on objective standards for classification.

Since the Moynihan Commission, the Deep State has actually invented new criteria that are not enshrined in law—terms like "law enforcement sensitive." The fact that something is considered law enforcement sensitive should not preclude Congress from performing its critical oversight role.

Another way we saw the Justice Department abuse this authority happened during the Clinton email investigation. Accused of mishandling information, Clinton had a certain number of emails that were classified—most after the fact, but some at the time she sent them. So guess what they do? They treat the whole investigation like it's a classified matter.

What do they get out of that? Secrecy. By classifying the whole investigation as top secret, it's really hard for Congress to shine a light on these things. It's all presumptively classified, even though there were a lot of documents that would never have been able to meet any objective classification criteria on their own. All of these concerns could be addressed by a new bipartisan classification commission.

Reconsider the Grand Jury Secrecy Rule

Another hiding place the Department of Justice uses to keep documents from ever seeing the light of day is a federal grand jury.

This secretive court system has been the target of reforms for decades, but little has changed to the eight-hundred-year-old concept of the secret jury.

According to Rule 6(e) of the Federal Rules of Criminal Procedure, evidence used in a grand jury proceeding cannot be made public. Conveniently for the Deep State, evidence once included as part of grand jury cannot be turned over to Congress.

Technically what the secrecy rule protects is the fact that a specific document was presented. It doesn't actually protect a document in and of itself from being disclosed. What we experienced with every Fast and Furious case was a DOJ that used this rule to its own advantage. If a case was going before a grand jury, the DOJ would code every document as 6(e)-protected because, in their view, any document they considered to be part of the case file may be shown to the grand jury. By designating the documents on the front end, they could avoid disclosing them if Congress came calling. In other words, instead of determining what documents were actually shown to a grand jury, they made all related documents off-limits to Congress.

In any package of reforms that addresses the grand jury process, we should consider whether the secrecy rule serves the best interests of the American people.

Information Accessibility

In response to well-deserved criticism about the FOIA request backlog, Obama administration officials often cited the increasing number of requests and the lack of budget to pay people to respond to those requests.

Any solution to the government's transparency issues must involve time-saving digital record keeping and cloud-based storage that could make routine reports easily accessible to reporters and the public. While this seems like a no-brainer, a surprising number of agencies have resisted making information so easily available.

Instead some agencies continue to use labor intensive paper-based systems while simultaneously complaining of a lack of manpower. One of the worst offenders was the Clinton State Department. Hillary Clinton demonstrated the technique when she left the department in 2013, turning in fifty-five thousand pages of printed emails on paper. Never mind the fact that State should never have had to go looking for her emails in the first place—the whole point of submitting paper copies was to slow the document production.

Uniform Classification

One of the great challenges for budget hawks who monitor government spending is the lack of uniformity in the way expenditures are tracked and recorded. This inconsistency serves the Deep State well by making apples-to-apples comparisons across agencies difficult.

The government, via the Office of Management and Budget, needs to implement a standardized, universal coding process that can be understood by the public and outside groups.

Without getting too wonky, each of the more than 320 federal departments and agencies operates with codes to match its budget line items with how the money is spent, awarded, or granted. But there is no universal standard for coding.

Imagine if there were no standard for transmitting electricity. If every time you went to plug an appliance into the wall and there was a different range of volts or wattage, sometimes your appliance would work and sometimes you would start a fire. Obviously that would be problematic.

The same is true with how the federal government operates.

In the medical insurance industry, one of the costs is how patient ailments are coded. Part of the reason you have so much administration of your doctor office visit is the need for the medical community to code everything. There is a code for just about everything. There is a specific code for an alligator bite and a dozen different codes for different types of insect bites.

It is good to track and understand trends, but the medical coding has become cumbersome and unwieldy.

Yet in today's electronic age there is no reason for the federal government not to adopt universal coding of expenditures, awards, and grants. In the name of transparency, the government should be making these expenditures available online so if you are interested you can review it.

The Sunlight Foundation pitched the project to the Trump administration this way:

The Trump administration should keep, strengthen, and seek to codify the open data policies that the Obama administration implemented, including the Open Government Directive, order on machine-readable data and the Digital Government Strategy. Public information should be posted online, data should be available in bulk and in structured formats, and systems that power open data should be supported and expanded. As we told the White House in 2017, anxiety about open data in the Trump administration has created doubt and uncertainty in many parts of American society, particularly in the scientific and academic communities. Businesses and entrepreneurs need to be able to trust that data disclosures will remain in place if they're going to be able to rely upon them. Those same stakeholders are negatively affected if the quality or periodicity of government data is diminished, as occurred in Canada. If the goal is to stimulate the use of open government data, it's important for a chief executive and cabinet secretaries to be cheerleaders for the quality of official statistics, not to cast doubt upon them. Increased risk discourages investment.

I truly hope our president and our Congress will take action before it is too late. We need to be more awake than ever. Because the Deep State does not sleep.

Epilogue

The task ahead of us is not an insurmountable one. The notion of a Congress checking the power of the executive branch is not new, nor is it revolutionary. In fact, believe it or not, the effort to check Congress goes all the way back to Francis Scott Key and Samuel Houston. The high-profile case in 1832 involved big names, an act of violence, and bureaucratic attempts to take on governing authority.

Francis Scott Key, author of "The Star-Spangled Banner" lyrics, and Samuel Houston, first president of the Republic of Texas, were the primary figures in one of the earliest efforts to check the power of the state.

The story starts with a federal contract. Houston lived with Cherokee soldiers as a boy, then became a soldier for Andrew Jackson and, subsequently, the governor of Tennessee. He suddenly quit that job when his wife, Eliza Allen, left him. That happened barely three months after their wedding. Exactly why, we don't know. It was rumored that he was an alcoholic, that she was in love with another man, among several other salacious details. It was that big a story at the time. What we do know for sure is that after a weeklong drinking bender, Houston went to live with a Cherokee tribe in Arkansas.

In 1830 he went to Washington, D.C., to bid on a contract that would supply food to Indians being forcibly moved west of the Mississippi River. He wasn't the lowest bidder, though. For a short time it looked like he would get the contract anyway even though that would have been illegal; his old army boss and mentor, Andrew Jackson, was now president of the United States and wanted him to have it. Press outcry over the near-scandal forced Jackson to abandon the idea and Houston returned to the Cherokees.

When Houston came back to D.C. two years later, he found a different political climate. Jackson's Democrat administration was under fire from the opposition Whig party. John Eaton had resigned as secretary of war; Eaton was the cabinet secretary who had nearly awarded Houston the contract.

In an anti-Jackson speech before the House on March 31, 1832, Ohio representative William Stanbery railed, "Was not the late Secretary of War removed because of his attempt fraudulently to give Governor Houston the contract for Indian rations?"

Houston heard what Stanbery said about him but couldn't do anything because representatives face no repercussions for remarks they made on the floor of the House or Senate. They could say whatever they wanted about a citizen and were immune from being sued. But when Houston read the remarks reprinted in a newspaper, he sprang into action.

First he sent Stanbery a letter essentially asking if what he read was true. The formality was required as part of the etiquette for fighting a duel, which historians say was Houston's intention. But when Stanbery wouldn't answer him, Houston had to fight back another way.

According to an account of the incident in *Tennessee Historical Quarterly*, on April 13, 1832, Houston was passing the evening with several congressmen in a hotel room. Upon leaving, he saw Stanbery across the street, approached him, and asked, "Are you Mr. Stanbery?"

"Yes sir," Stanbery replied, not recognizing Houston.

"Then you are a damned rascal," exclaimed Houston, cracking the Ohioan over the head with his hickory cane.

The assault continued. At one point Stanbery pulled out a pistol and tried to fire it into Houston's chest, but the gun didn't go off. "Houston tore the weapon from Stanbery's hand and went on with his caning until the agonized man broke down and whimpered."

The following day Stanbery sent a note to Andrew Stevenson, the Speaker of the House, detailing the beating. Houston was arrested, arraigned, and made to appear before the House clad in his "fur-collared buckskin coat and carrying the now famous cane of Hermitage hickory," says the article. With him was his lawyer, Francis Scott Key.

The trial began on April 19, 1832, and lasted several days. Highlights included Stanbery admitting that he didn't mean to say Houston committed fraud, witnesses being called liars, President Jackson throwing a bag of coins at Houston and telling him to "dress like a gentleman," someone dropping a bouquet of flowers at Houston's feet after his opening remarks, and the defendant's final passionate speech in his own defense wherein he declared, "Sir, when you shall have destroyed the pride of American character, you will have destroyed the brightest jewel that heaven ever made."

The *Tennessee Historical Quarterly* piece dryly noted, "Whether Francis Scott Key, who sat in the front row, felt like disowning certain lines of his own, inspired by the bombardment of Fort McHenry, is a detail upon which history is remiss."

In the end, Houston was found guilty but received only a reprimand from the House. Stanbery filed a complaint in civil court and Houston was fined five hundred dollars.

He never had to pay it because President Jackson set aside the sentence: "Though the principle of Congressional immunity and the right of Congress to find private citizens in contempt was upheld, doubt had been raised as to whether character assassination should be covered by that immunity," the *Quarterly* tells us.

So if you ever wondered how former Democrat Senate majority leader Harry Reid got away with saying Governor Mitt Romney never paid taxes—a lie he later admitted he made up—or assaulted the patriotism of businessmen Charles and David Koch 130 times on the Senate floor . . . that's how. Immunity.

Changing Course

As I watched all of this play out from my vantage point in the U.S. House of Representatives, I came to realize where the real power resides. It's not in the halls of Congress. It's not in the Oval Office. It's not even in the darkened corridors of Deep State power.

The real power resides with the people of the United States of America. Nothing will change unless we demand it. But if we demand it, the Deep State is helpless to override us.

As powerful as the position of House Oversight Committee chairman is, it's not powerful enough to change hearts and minds. I realized I could accomplish more on the airwaves than I ever could on the dais of the House Oversight Committee or on the floor of the U.S. House of Representatives.

I wrote this book because we can no longer stand by and let the bureaucrats, the politicians, and the big businesses collude against us. Our power must be asserted in all three branches of government. We've allowed the Congress, which represents us more directly than any other branch, to be usurped by unaccountable bureaucrats, self-interested partisans, and profit-hungry business interests.

My call to action is this: This election cycle, ask your candidates for Congress what they will do to restore the checks and balances we need. What will they do to assert the will of the people over the will of the bureaucrats? Where do they stand on impeaching executive branch officials? Would the candidates support civil service reforms? Do they know what inherent contempt power is and would they be willing to reclaim it?

These are not simple issues, but they are important ones. We need a government with three fully functional branches. The American people are the only ones who can give it to us.

Acknowledgments

"All my best ideas I got from someone else" wrote Sam Walton. So true! The people around me have shaped me, helped me, and encouraged me to do things I would never be able to do otherwise.

When I was growing up my parents always encouraged my brother, Alex, and me. They presented opportunities where certainly we would fail, but perhaps we would succeed. They projected an attitude of "If not you, who?" We weren't the biggest, the fastest, the smartest, the tallest, the shortest, or different enough to stand out from the crowd in any one area. But what did differentiate us was attitude and perseverance.

I like to tell my kids, "Attitude is everything." It is often the tipping point between failure and success. From the youngest of years, I distinctly remember knowing it is okay to fall, but keep getting up, dust yourself off, and keep moving forward. "You can't win if you don't play" was the prevailing attitude in the Chaffetz house.

As I reflect upon my life, I see there were always people willing to go out of their way to help me, to encourage me, to spend time with me, and to help me achieve important goals. From my teachers to my soccer coaches to my business colleagues, friends, and family, I have been fortunate to be molded and shaped by all these people.

My parents passed away a number of years ago, both from

cancer. I miss them. I wish they were still here to see the publication of this book. I am sure they would be pleased. They were proud of everything I did—even the lamp I made in shop class that almost burned down our house.

By my side since age three has been my brother, Alex. He means the world to me. I love him and our bond is unique and precious to me. He has often been my political muse, my check on reality, and my source of humility if things got out of hand. I cherish our daily interaction.

Certainly this book would not have been possible without first getting to Congress. I am very grateful to Jon Huntsman Jr. and his entire family for an opportunity to join his gubernatorial bid to become Utah's sixteenth governor. He won and took me along for the ride, which opened up so many other possibilities.

From the start of my own run, Jennifer Scott has always been by my side with her insight, savvy, talents, candor, and hard work. She was my first campaign manager, a loyal member of my staff, and was critical to the writing of this book. Jennifer is my secret weapon and a pivotal part of my success. I can't thank her enough.

Through the years I had numerous people willing to step up and help a fledgling campaign succeed. In our Congressional office, we were able to assemble amazing talent to serve the people of Utah's Third Congressional District. We had an amazing ride together. I am grateful to each person who served on our staff.

Despite its reputation, Congress comprises lots of really good people; not all of them, but most of them. For those that have served, thank you for your time and sacrifices. Serving with you was an honor and a privilege.

Writing a book was new but I was fortunate to have Elaine Lafferty to guide me. She is an exceptional talent who brought her know-how to writing and compiling *The Deep State*. It could not have happened without her skills and touch. Thank you!

Eric Nelson from Broadside/HarperCollins had the vision and perfect approach to coach a rookie author. With a world of options in front of him, he believed in us. Eric Meyers provided critical administrative help. I appreciate his expertise and professional approach. I can't thank them and our publisher enough for turning this book into a reality.

David Larabell from Creative Artists Agency (CAA) had the belief in me and the ability to put together a win-win situation for all. This man has skills!

I am also very thankful to Rupert Murdoch and the Fox News family. After thirteen years in politics with eight and half years in the U.S. Congress, I wanted to apply my perspective and share my insights. Fox News was a perfect fit. As I started day one at Fox News, I asked what they wanted from me. "Just you be you. Be authentic and tell us what you really think." That is all they ever asked for, and I am humbled to be a contributor to such an important organization.

But what matters most to me, what has the greatest impact, and where I am truly blessed, is my family.

My wife, Julie, is the sweetest person I have ever met. My love for her and our children is the most important part of my life. I could not do what I do, or achieve what I have been able to achieve, without her and our wonderful, talented, understanding children. Being in politics is a crazy business, but the cohesive part of our family is Julie. Without her I would be lost. Thank you, Julie, from every ounce of my body. Thank you!

And to our kids, thank you for your never-ending patience. I was gone more than I was home when I was in Congress. I missed a lot, but most of all I missed the time with each of you.

You always said you understood, but you will never know how deeply my heart ached being away night after night.

I hope you enjoy the book as much I have had putting it together.

God bless,
Jason Chaffetz

Index

About the Author

JASON CHAFFETZ is an American politician and Fox News contributor. He served as a U.S. representative from Utah for eight years, including time as the chairman of the House Committee on Oversight and Government Reform from 2015 until 2017.